# · DENNIS KING

# GET
# THE FACTS
# ON ANYONE

PRENTICE HALL

New York London Toronto Sydney Tokyo Singapore

Prentice Hall General Reference
15 Columbus Circle
New York, NY 10023

An Arco Book

Arco, Prentice Hall and colophons are
registered trademarks of Simon & Schuster, Inc.

Manufactured in the United States of America

1  2  3  4  5  6  7  8  9  10

**Library of Congress Cataloging-in-Publication Data**

King, Dennis
    Get the facts on anyone / Dennis King.
    p.  cm.
    ISBN 0-13-351859-0
    1. Public records—United States—States—Handbooks, manuals, etc.
2. Biography—Research—Methodology—Handbooks, manuals, etc.
3. Investigations—Handbooks, manuals, etc. I. Title.
JK2445.P82K55     1992                                   91-40513
353.0071'4'0202—dc20                                         CIP

# CONTENTS

## 4 • Finding "Missing" People  35

## 5 • Backgrounding the Individual—Biographical Reference Works  68

## 9 • Court Records   122

## 10 • Backgrounding the Individual—Miscellaneous Records and Resources   135

# ACKNOWLEDGEMENTS

This book could never have been completed without the help of Geraldine Pauling, Kalev Pehme, Katy Morgan, Dave Phillips, Chip Berlet, A.J. Weberman, and Steve Weinberg; the staff of New York University's Tamiment Institute Library; the New York Public Library and Brooklyn Public Library telephone reference services; my agent Nancy Love; and my editor at Arco Books, Barbara Gilson.

# INTRODUCTION

The purpose of this manual is to assist researchers in compiling accurate background or profile information on individuals, business entities, and nonprofit organizations. It can be used as a "where's what" guide for finding the answers to relatively simple questions, or as a manual for comprehensive (deep background) investigations.

The manual is organized in a cumulative manner, proceeding from nuts and bolts techniques (e.g., locating a person whose address is unknown) to the backgrounding of individuals and then to more complicated research tasks. You will find that methods mastered at one stage not only retain, but increase their usefulness at later stages.

County courthouse and state government records are a major focus of this book. It is not easy to generalize about these records, because of the variations in filing systems, laws, and administrative policies affecting the public's right to know. Sometimes the descriptions are based on my own experience with New York City records. In other instances, I rely on what I believe to be the most common system. To avoid oversimplification, I make frequent use of "may," "might," "sometimes," and "often." I try to provide several alternative methods for gaining each type of information, leaving it to you to select the way that best fits your investigative requirements.

While attempting to meet the needs of researchers of all types, this book includes special tactics for journalists and public-interest researchers who lack the access to confidential government records enjoyed by law enforcement officers, and who wish to avoid the kinds of trickery employed by collection agency skip tracers and private investigators. Although I describe a few typical ruses (they are, for better or worse, part of the real world of investigating), I also try to show that a researcher who exercises his or her ingenuity can usually find an alternate path to the same information or else an alternate body of equally useful information about the person or entity under investigation.

This is the first edition of a manual that I hope will go through many editions. I urge readers to send their suggestions and criticisms to me care of the publisher.

# 1 ·

# Basic Concepts

## 1.1 The Paper Trail

We live in a record-keeping society. Millions of Americans work in white-collar jobs involving creation, storage, and dissemination of data for government, business, or private institutions. The computerization of this function since the late 1960s has produced vast changes in research techniques in every field. The great turn-of-the-century pioneer of investigative journalism, Lincoln Steffens, would be awestruck by the resources that have replaced the ledger books and filing cabinets of his day.

Years ago the term "paper trail" was coined to refer to the vast wealth of records accumulated about an individual during his or her lifetime. Today the trail of paper has largely become a trail of computer bytes, yet the underlying concept is more valid than ever: It is almost impossible for anyone in our society to avoid leaving a trail of personal information in the files of government and private institutions. These documents provide a record of virtually every major event in a person's life: birth, baptism, high school and/or college graduation, military service, marriage, births of children, purchase of a home, deaths of parents, movement from one job to another, major illnesses, retirement, death. Also on record will be divorces, personal bankruptcy filings, criminal convictions, judgments obtained against subject in civil cases (with any liens or wage garnishments resulting therefrom), and even a list of subject's unpaid parking tickets.

By following the paper trail, you can study the influences prior to subject's birth that helped to mold his or her life—the backgrounds of both parents, their marriage(s), the births of older siblings, the family's genealogical records going back generations. You can also follow your subject beyond the grave—by going to the Probate Court to find out what happened to his or her estate.

Records compiled by utility companies, banks, and credit card vendors will also be part of subject's paper trail. Usually kept for limited periods only, these records will include lists of every phone number dialed from subject's home or office phone, every deposit made into (or every check drawn upon) his or her bank account, and every credit card transaction. Although such information is supposed to be confidential, private investigators with the right connections routinely gain access.

Subject's paper trail may include dozens of news articles about his or her activities. A budding investigator should therefore learn how to access these articles through newspaper and periodical databases, clippings "morgues," and the microfilm collection at the local public library.

Corporations and nonprofit organizations also leave a paper trail. Like an individual, a corporation has its "birth certificate" (certificate of incorporation), its "marriage certificate" (merger papers), and its major and minor crises (lawsuits, bankruptcy proceedings). It may even have a "death certificate" (certificate of dissolution). National and local business periodicals often report on such events as assiduously as the tabloid press reports on the escapades of movie stars.

## 1.2 The People Trail

The aim of following the paper trail is not simply to accumulate as many documents as possible. Although documents are important in their own right, they are also useful because they lead you to live sources: first, people with direct personal knowledge of your subject; second, experts with background knowledge who can steer you to the direct sources and can also help you to interpret what you find.

In backgrounding an individual, you might seek out his or her former neighbors, co-workers, or business associates. In backgrounding a corporation, you might contact its customers, suppliers, stockholders, or former employees. In backgrounding either an individual or a corporation, you would want to talk to their adversaries in any lawsuits. You would also want to talk to someone with a rosier viewpoint: the individual's best friend or the corporation's public relations consultant.

Success in any investigation depends on the skillful interweaving of paper trail and people trail. The paper trail leads to people with special knowledge who in turn steer you to new documentation, which then leads to people with even more (and hopefully deeper) knowledge. This spiral process, from documents to people and back again, gradually leads you to the heart of the investigation, possibly even to the proverbial smoking gun.

## 1.3 Parallel Backgrounding

If your subject is closely linked to a particular business or organization, the latter will have its own paper/people trail. By following it, you may obtain information about subject unavailable from his or her personal records. For instance, the personal records on Mr. X may contain no negative information, but the city housing authority's files on his contracting firm may contain documents suggesting that this supposed solid citizen is involved in rigging bids.

The same principle also works in reverse: If your main target is a business enterprise or nonprofit organization, you may gain startling insights by examining the personal backgrounds of its principals or officers. That seemingly innocuous annual report of your local community development corporation may appear in a different light once you learn that that the city has padlocked two buildings owned by the executive director because of illegal gambling on the premises.

Parallel backgrounding also may involve looking into the affairs of one or more of Mr. Y's business associates, relatives, etc., to gain information about or insight into Mr. Y himself (a classic example of this was the media's focus in the early 1970s on Richard Nixon's close friend Bebe Rebozo). Or, to gain insight into Corporation Z's business tactics, you might take a look at its chief rival (especially if the latter's business methods are better documented than Z's in lawsuits and government enforcement proceedings).

We are speaking of four basic types of parallel backgrounding: Personal/Personal; Personal/Corporate; Corporate/Personal; Corporate/Corporate. If you are beginning a complicated investigation, you might draw up a chart with each of these headings. As you accumulate names and other information, jot down possible leads under each heading. You probably won't have time to follow up more than a few, but the chart will help you to decide priorities.

## 1.4 Indirect Backgrounding

Essentially, this method is parallel backgrounding on a grand scale. You may find that your subject is linked in complex ways to various economic or political interests. The only way to understand the significance of the relationships involved—and to identify which, if any, of the individuals or organizations warrant parallel backgrounding—is to analyze this larger environment. This approach can sometimes lure you into unproductive areas, but it can also pay big dividends. In one recent case, background research on the economy of a West African nation to which a New York businessman had often traveled, helped to identify possible Libyan connections of that businessman. In another case, inquiries into the history of the Teamsters Union and of certain Midwest organized crime families led to a major breakthrough in understanding neofascist leader Lyndon LaRouche's links to the underworld.

This method, like direct backgrounding, involves both a paper trail (in this instance, books, newspaper and periodical articles, and various library and filing cabinet gleanings) and a people trail (chiefly, the "experts").

## 1.5 Operative Backgrounding

This is the level at which you put everything together. Operative backgrounding is the process of figuring out how things work in a particular area of money and power, and then interpreting the facts in light of that understanding. To understand a city politician, you have to understand the world in which he or she moves—the relationships between the politicians and established wealth on the one hand, and between the politicians and organized crime on the other. You have to understand the mechanisms of legal and illegal graft through which transactions between these three forces are conducted. Likewise, to understand a local hoodlum you have to know how organized crime works—its division into so-called crime families, the characteristic businesses these families get into (and why), how they "launder" their illegal income, and how they deal with both the politicians and the police. The principle also applies to my own specialty: the study of cults and extremist groups. Here you enter a world where greed and the desire for power and status are covered up by high-minded ideologies (or theologies) that must be decoded to discover the underlying interests and the real meaning of the incessant factionalism (often just an inverted form of capitalist competition taking place in a frog pond with status rather than cash as the payoff).

As this book is not a political treatise, I have only dealt with operational backgrounding when necessary to explain specialized areas of research. But the best achievements in investigative journalism usually are a result of having an understanding of these matters—this is what guides the journalist almost uncannily to the right sources. You will not always learn much about this from academic social scientists, who frequently either prettify things or ideologize them, and thus miss the main point. For a more pragmatic view (including a healthy cynicism about human nature), seek out veteran political and crime reporters. Better yet, cultivate insiders in the worlds of business and politics. It is not easy to get CEO's to open up, but you can always find someone on a lower echelon—or a retired or fired executive or an independent consultant—who knows as much if not more about the way things work and is willing to talk frankly.

# 2.

# Some Basic Research Tools and Resources

## 2.1 Your Office Reference Shelf

The following is a description of the various directories, manuals, periodicals, and other reference tools that are most useful in investigative research.

- Current telephone directories. You should have the current white-page and yellow-page directories for your entire metropolitan area, and the directories for Manhattan, Washington D.C., and your state capital. To save money, keep beside your phone the *AT&T Toll-Free 800 Directory*. Its two volumes (consumer and business) include a total of 180,000 listings.

- Back-issue phone directories. When a current phone directory is supplanted, don't throw it out. Back issues are quite useful in finding people not listed in the current edition (see 4.5 and 4.8).

- *The National Directory of Addresses and Telephone Numbers*. This immensely useful annual softcover volume, published by General Information, Inc., covers all fifty states and contains 110,000 alphabetical listings as well as a classified section with hundreds of business, nonprofit, and government subsections. For an investigator, the most helpful features are the state by state listings for county governments, state government agencies, and public libraries. Also see the listings for regional and local offices of every federal bureau, agency, and commission, and the listings by subcategory for hundreds of newsletters and other specialty periodicals.

- Local city directory or crisscross directory (see 2.2 and 2.3). These can be expensive. Only subscribe if you do a lot of tracing of hard-to-find people; otherwise, use the copy at your local public library or chamber of commerce, or else consult an online crisscross database via an information broker (see 2.6).

- Guidebooks to directories. The big name in this field is *Directories in Print* (formerly *Directory of Directories*). This is an annual, two-volume reference published by Gale Research Inc. The eighth edition describes almost 14,000 directories arranged under twenty-six categories; another 1,000 are listed in a supplement. *Directories in Print* includes only directories that are national or regional in scope. It should be used in tandem with *City & State Directories in Print*, which covers 5,000 localized directories, i.e., directories for tri-state or smaller regions. (Note that the latter book will be merged into *Directories in Print* in 1992.) Also to be consulted is the *Guide to American Directories*, published by B. Klein. The richness of information available in these mega-directories is extraordinary. Of all reference works, I have found them the most useful. If you consult them at the beginning of your investigation, before you traipse to the public library, you can work out an efficient plan of what directories to consult and in what order.

- Back editions of reference directories. Most directories found in public libraries (e.g., *Martindale-Hubble Law Directory*) are too expensive for freelance journalists and other shoestring investigators to purchase for their offices. However, many libraries will throw out an old directory or offer it for sale at a nominal price as soon as the new edition hits the shelf. If you become friendly with your local reference librarian (as any good investigator should), you can learn when a book you need is about to be discarded. Note that in many investigations the back editions of a directory are more useful than the current one.

- State and local government directories. Most states and large cities publish annual volumes giving the addresses and phone numbers— and sometimes the functions—of all state or local government agencies and legislative committees. These volumes also give the names and phone numbers of legislators, legislative committee staffers, and key administrative officials. Most important, they contain lists of all professional and commercial licenses required by the state or city government, and the agency responsible for each license.

- Directories and guides to the federal government. The *United States Government Manual*, published by the Government Printing Office, outlines the organizational structure, functions, and key personnel of each federal department or agency. The *Congressional Directory* describes the various congressional resources, including committee and subcommittee research staffs. You should also have *Washington Information Directory*, which is indexed by subject and government department or agency (and also includes nongovernment Washington sources); and *Lesko's Info-Power* (formerly *Information U.S.A.*), which describes how to squeeze free information out of the bureaucrats. Juggling these four books, you can usually figure out which bureaucrat or congressional aide is most likely to have access to the information you need, and which department or agency is most likely to have public records (or records accessible under the Freedom of Information Act) relevant to your research.

- Guidebooks to local public records. In-depth "where's what" directories to municipal, county, and/or state records have been compiled (either in published or unpublished form) by college journalism departments, daily newspapers, or public-interest groups in various localities. Check with the librarian of your local newspaper—or with any nearby college journalism department—to see if there is a manual for your city or state. If a local newspaper has produced a private manual for its staff reporters, request a courtesy copy. A list of several of these "where's what" manuals is contained in the bibliography. Since the same kinds of public records are kept in every locality, albeit in different formats and under varying restrictions, a manual for one locality will be useful in another.

  If you intend to do much investigating outside your own city or state, invest in *The Guide to Background Investigations* (described by journalism professors as the best compilation on state and local records) or in the *State Records Access Directory*. Through either of these books (see bibliography) you can learn not only where to write for particular records, but also which offices will provide information over the phone (crucial for any reporter on deadline!). Another important work is Privacy Journal's *Compilation of State and Federal Privacy Laws*, which describes more than 650 state and federal laws affecting privacy. By telling you what's not available (and what shouldn't be, but is), this resource will help you plan your public records' search strategy.

- Investigative how-to manuals. Several are listed in the bibliography. The most essential is *The Reporters' Handbook* (second edition), written and edited by members of Investigative Reporters and Editors (IRE), a nationwide professional organization with headquarters at the University of Missouri.

- Offbeat how-to manuals. You can learn all about the use of false ID, illegal electronic surveillance, computer hacking, money laundering, and similar arcane skills from scores of pamphlets offered by mail-order publishers. Many of these manuals are written by and for criminals, and rarely differentiate between what is legal and what can land you behind bars. Yet they contain much valuable information for an investigator. The most comprehensive mail-order catalog is offered by Loompanics Unlimited (see Appendix). Some of the rather expensive pamphlets in this and other catalogs simply repeat each other, so it's best to seek out a bookstore in your area catering to survivalists and *Soldier of Fortune* readers so you can examine before buying.

- Periodicals. For tips on the latest investigative techniques, subscribe to *The IRE Journal*, a bimonthly magazine published by Investigative Reporters and Editors. Without fail, you should order a full set of the back issues since 1978 (available at a very reasonable price) and the latest cumulative index (published in IRE's 1989 Membership Directory). Additional investigative tips can be gleaned from the monthly *Privacy Journal*; hardcopy and microfiche backfiles and an index covering the years 1974-1989 are available from its publisher. You might also subscribe to journals for professional librarians (e.g., *Special Libraries*), which are filled with advertisements for useful research products and services. (See Appendix for addresses of the above publications.)

- Conference proceedings. IRE holds annual national conferences with extraordinarily detailed panels on investigative techniques. Cassette tapes of every panel at every conference since 1980 are available from Sound Images Inc., P.O. Box 460519, Aurora, CO 80046. An index to these panels was published in the Winter 1989 *IRE Journal*. Note that private investigators' associations hold similar conferences and panel presentations; to see if these are likewise available on cassette or in published form, contact the relevant associations listed in the *Encyclopedia of Associations*.

- Reference book and microform publishers' catalogs. You can keep up with these rapidly changing fields by getting on the mailing lists for the major publishers' catalogs, supplements, and news releases. For a list of the most important publishers, see the Appendix.

- Database directories, database users' newsletters, and database vendors' catalogs (see 2.6 and the Appendix). Even if you don't own a computer, these will keep you up-to-date on what can be obtained from commercial or public library search services.

## 2.2 City Directories

Ever since the 19th century, specialized publishers have produced household-by-household and store-by-store marketing directories, commonly known as city directories. These books are used by telephone or door-to-door sales teams, direct mail firms, fundraising experts, pollsters, newspaper subscription departments, and essentially anyone who needs marketing information that will identify potential customers and/or supplement the demographic information found in U.S. Census reports. Such directories are also used by private investigators, skip tracers, police detectives, and newspaper reporters to locate individuals and compile background information on them. By using back as well as current editions of such directories, along with back and current phone books, you can gather a remarkable amount of information about someone in a short time.

The city directory is compiled via door-to-door or telephone surveys. It may tell how many people live in a household, how long they have been there, where the head of the household works (or at least, what his or her occupation is), the names of other household residents (spouse, roommate, children, etc.), the general income level of the neighborhood, and the household telephone number(s). By tracing a person's name through back issues of a city directory, you can get a bare-bones picture of his or her family through the years.

Usually revised each year, the city directories include alphabetical, street, and numerical listings. The street listings will help you to find subject's present and former neighbors, and also give you a sense of the general neighborhood environment.

Most large cities are no longer covered by city directories—urban mobility and the socioeconomic disintegration of inner-city life have rendered them impractical. However, city directories are often still published for the suburban communities surrounding the core cities. In addition, many medium-sized and smaller non-suburban cities are covered, as well as many small towns and rural areas (the latter, sometimes, by special rural route directories).

The current and back editions of a city directory may be found at the local public library or chamber of commerce. Among city directory publishers, the biggest name is R.L. Polk & Co. (about 1,300 directories in the United States and Canada). Collections of Polk directories are available at its five district offices.

In cities no longer covered by a city directory, the public library will have copies up through the final edition. A researcher can thus trace a longtime city resident's life up to that point. Furthermore, many persons listed in now-defunct city directories in the 1940s and 1950s later moved to suburban communities that are still covered. You can thus often compile an uninterrupted record of successive residences, household members, and neighbors. (Even if you lose the city directory trail, you can pick it up with back-edition crisscross and telephone directories; see 2.3 below and 4.5.)

## 2.3 Crisscross Directories

These directories are based not on survey information but on the information contained in the local telephone book. Unlike the city directory, the crisscross directory does not include alphabetical listings; it therefore must be used in tandem with the telephone book.

Essentially, the crisscross directory rearranges the phone directory. Instead of listing phone numbers alphabetically by customer's name, it lists them in numerical order and by street address (it is thus often called a "reverse directory"). When you have a number but no name, you look in the numerical listings and get both name and address. When you have an address but no name, you look in the street listings. Sometimes the street listings and the numerical listings are in one volume, sometimes separate. A volume including street listings is sometimes called a "street directory" or "household directory."

As noted above, crisscross directories cover the large cities abandoned by the city directories. Although providing less information than the latter, crisscross directories should never be underestimated as a source of information (especially if you use back as well as current editions). For instance, Cole's directories for the five boroughs of New York City will tell you how many years subject has been listed at his or her current address; whether subject's new listing is altogether new to the directory or only new for the given address; the identities of two or more people with different last names who are sharing a phone number listed separately under each name; the identities of two or more people with different last names who have separate phones at the same street address; the names and phone numbers of subject's neighbors; and the approximate income level ("wealth rating") of the block.

Cole's is one of the largest publishers of crisscross directories. It leases rather than sells them. Subscribers have telephone access to information from Cole's past and current directories nationwide via its subscriber's private line service.

Like city directories, crisscross directories are usually available at the local public library or chamber of commerce.

## 2.4 The Law Library

Whether you're investigating an individual, a corporation, a nonprofit organization, a trade union, or an electoral campaign committee, they are all subject to specific legal statutes, government regulations, judicial decisions, and administrative rulings. Hence, a law library (or the legal databases, LEXIS and WESTLAW) can provide important information about your subject. For instance, you can look up the laws and regulations relating to Mr. A's activities as a street peddler; Dr. B's, as a podiatrist; Ms. C's, as a stockbroker.

To guide your search for such information, there are four essential sets of books: the city code, the state code (for instance, *McKinney's Consolidated Laws of New York Annotated*), the *United States Code Annotated*, and the *Code of Federal Regulations*. The designation "annotated" means that a set provides, along with the text of each section of the law, a summary of the most important decisions interpreting it. Each annotated volume, unless it is from the latest annual edition, will include a "pocket part," an annual update inserted in a pocket at the rear of the volume. The pocket part gives all new developments since the date of publication of the volume on the shelf and should *always* be consulted. Note that the *Code of Federal Regulations* does not have pocket parts; you must consult the *Federal Register* for the latest developments.

Whichever code you are looking at, it will be indexed according to topics/key words in an easy-to-use manner. Indeed, the index itself may consist of several volumes. With the index as your guide, your use of these volumes is only limited by your ingenuity and your knowledge of subject's activities. Mr. X is a cafe owner? Look at the municipal laws relating to eating establishments, including the health, fire, and sidewalk codes; also look at the city and state laws pertaining to registration of small businesses. Ms. B is a freelance writer? Look at the state and federal tax codes and regulations pertaining to self-employed individuals who file itemized deductions.

If a person is engaged in a licensed occupation, the state code may guide you to a surprising array of official records (see 10.11 and 10.12).

On the federal level, you might want to skip the *United States Code Annotated* and go straight to the *Code of Federal Regulations*, which includes a volume called the *CFR Index and Finding Aids*. Let's say you are investigating Mr. W., a right-wing arms dealer suspected of supplying machine guns to the Ku Klux Klan. Look under "Arms and munitions," and note the relevant subtopics, cited by "title" and "part." Turn to Title 27, Part 178 ("Commerce in Firearms and Ammunition") and Part 179 ("Machine Guns, Destructive Devices, and Certain Other Firearms"). Here you will find a description of the various filing requirements with which Mr. W. must comply. Your next step: Check with the Alcohol, Tobacco, and Firearms Bureau to find out which of the government forms filed by Mr. W. are available under the Freedom of Information Act.

## 2.5 Freedom of Information Laws

For generations, bureaucrats routinely denied the public access to most records of the federal government's executive arm. Congress initiated a new "open government" approach in 1966 by passing the Freedom of Information Act (FOIA). This law, amended in 1976 and 1986, applies to all departments and agencies of the executive branch, including the Armed Forces and the CIA, but it excludes Congress, the federal court system, and the president's immediate staff. Essentially, the law says that the bureaucrats and brass must provide copies, to anyone who requests them (including a mobster in prison), of any government document except those covered by nine exemptions. Exempt documents (or exempt portions of documents) include classified national-security information, trade secrets and other confidential business information, information the release of which would violate personal privacy, information about ongoing law enforcement investigations, information that might jeopardize a law enforcement informant, and certain internal bureaucratic memoranda.

The exemptions may seem to provide loopholes for the bureaucrats to weasel out of giving you just about anything. In fact, a vast amount of material is readily available to anyone who bothers to request it. America's corporations and their foreign competitors use the FOIA assiduously to gather government documents that will give them a business edge. Journalists use the FOIA in preparing scoops that blast the very agency releasing the information (if there hasn't been more of this, it's because most journalists are too lazy to master this tool). Public-interest foundations use the FOIA to gather large libraries of national security documents that illuminate every conspiracy and intrigue of the Cold War years. Former radicals have used it to gather their own files from the FBI, and then have turned around and successfully sued the FBI. I myself have used it to gain information from the FBI, the CIA, the Department of Energy, and the State Department. In each case I found that the particular department or agency's FOIA staff complied with the spirit as well as the letter of the law, even though the material released was potentially embarrassing to the government. My experiences may not have been typical—many of my fellow journalists have complained of bureaucratic stonewalling. Nevertheless, a 1985 congressional study found that over 90 percent of FOIA requests were being complied with adequately.

How can you use the FOIA to background a local businessman or mobster? Because of the Privacy Act, you can't just expect a government department or agency to send you everything they have on someone. (Nor should you make requests for information that is clearly personal—the FOIA unit may contact the person pursuant to the Privacy Act and tell him or her about your request.) However, documents pertaining to businesses, nonprofit organizations, government contracts, etc. with which subject is associated will be available. You thus can do parallel and indirect backgrounding (see 1.3 and 1.4) on a broad scale, gaining much information about subject in the process.

Let's say you need information about Arthur, a community development corporation director in Chicago who has wangled tens of millions of dollars from the federal government to finance development projects for the black community (but has only a collection of almost bankrupt enterprises to show for it). Under the FOIA, you can obtain the relevant files of the succession of federal agencies that gave him the money—these files will include much of the correspondence and many of the intra-agency memos that led up to each grant. (If the FOIA officer is really conscientious, you may even receive a copy of the letter from a U.S. Senator in Arthur's state supporting Arthur's request for yet more money.) You also can get the audits and the records of any resulting investigations. You can see who in the agency pushed for the investigations and who, higher up, apparently quashed them.

Further, you can look at the HUD records on a housing project financed by Arthur's organization. You can look at the FDIC's files on the community savings bank controlled by Arthur—files that may include devastating criticism of the bank management selected by Arthur and his cronies. You can get the FBI's file on Arthur's late bodyguard (a former Black Panther) who died in the mysterious crash of a plane owned by Arthur; the FAA report on that crash; license information on the pilot who died along with the bodyguard; perhaps even DEA files regarding the mysterious airstrip from which they had taken off. And as you collect all these documents you will automatically be gaining the names of potential sources—those people in government, formerly in government, or outside government who opposed giving money to Arthur or tried to blow the whistle on him.

The main problem with the FOIA is the time it takes to get an answer. Although the government is supposed to reply to any request within ten days, that reply is simply an acknowledgment that the request has been received and that it will be processed in its turn. During the Reagan years the staffs of FOIA units were deliberately cut back, creating backlogs at some departments (especially the FBI). However, departments with a low volume of requests and little need to redact documents for national security reasons often still meet requests with reasonable promptness.

The FOIA is best used in investigations that are not run on a tight deadline. But even if you have as much time as you like to gather the story, you should make your FOIA requests as soon as possible—the documents you receive may open up an entirely new avenue of inquiry.

The following suggestions are aimed at getting the maximum information with the minimum wait:

- Do your homework. Each agency is required under the Privacy Act to publish annually a description of its records systems and the categories of individuals on whom records are kept. These notices can be found in the *Federal Register* (computer-searchable via DIALOG; see 2.6) or you can obtain a copy from the given agency's FOIA unit. The biennial *Privacy Act Issuances* is a compilation of these notices from every agency covered by the act. Much of the same information is contained in the *Code of Federal Regulations*. Also useful in figuring out what's in an agency's files is the Office of Management and Budget's monthly agency-by-agency inventory of every red-tape form and procedure by which information is gathered from the public. Printouts of the inventory for a given agency can be obtained from that agency. To learn about defunct forms and procedures, request back copies of the inventory.

- Contact the press office of the agency in question. In some instances, if you tell them you are on deadline, they will get you the information directly without your having to go through the FOIA procedure. This is easiest to do if you have a very simple request; e.g., for a single document you already know exists.

- Get your congressperson or U.S. senator to obtain the information for you. He or she can go through the particular department or agency's congressional liaison office, and often get the records you need within days. If your senator or congressperson won't help you (or is newly elected and lacking in clout), try a member of Congress who has a special interest in the issue you are researching.

- Make sure the information you need really requires an FOIA request. I once asked the Federal Election Commission for information, under the FOIA, that was routinely available without an FOIA request. The press officer called me to gently suggest I withdraw my request so he could send me the information immediately. Not all government agencies will volunteer such advice.

- Check to make sure the documents you've requested have not already been released. Each department or agency covered by the FOIA keeps an index of released documents. If what you want or part of what you want is on the index, you can order copies directly from the agency's library with no delay. (Note, however, that some departments or agencies do not have very good indexing systems; the best index is that of the FBI.)

- Phrase your request clearly and be as specific as possible. If your request is vague or overly broad, the bureaucrats may use this as an excuse to deny it altogether. For detailed information on how to prepare a request, see "How to Use the Federal FOI Act," a pamphlet published by the Reporters Committee for Freedom of the Press. Also see "A Citizen's Guide on Using the Freedom of Information Act and the Privacy Act of 1974 to Request Government Records," a somewhat longer booklet available from the House Committee on Government Operations.

- When you make your request, get someone else to ask for the same information in a slightly different form. If the requests are processed by two different FOIA officers within the same unit, one may release things that the other withholds, and vice versa.

- Try more than one agency. Copies of memos from Agency X will often end up in the files of Agency Y as well. Agency X might regard the memos as too embarrassing to release; Agency Y may release them without hesitation.

- File your request with an agency's field offices and regional offices as well as its Washington headquarters; in some agencies, these units make their own determinations about FOIA requests.

- Think creatively. James Bamford, author of *The Puzzle Palace*, wanted to know the number of employees at the top secret National Security Agency (the electronic surveillance and code-breaking agency). The NSA stonewalled him on this and everything else. He eventually obtained the number of employees by making an FOIA request to the

U.S. National Credit Union Administration for its records on the Tower Federal Credit Union, which is located at the NSA. He also used a loophole in the FOIA to obtain copies of the NSA's internal newsletter.

- If an agency gives you part of what you want, but withholds more, file an immediate appeal and also let the bureacrats know you'll take the matter to court if necessary. Get your congressman to write a letter on your behalf. Alert the chairmen of the House and Senate FOIA oversight committees (in the Senate this is the Subcommittee on Technology and the Law; in the House, it's the Subcommittee on Government Information, Justice, and Agriculture). Almost always, the bureaucrats will release a few more items to avoid a hassle.

- Seek help from experts. Journalists can contact the Reporters Committee for Freedom of the Press, Suite 504, 1735 I Street, N.W., Washington, D.C. 20006. Nonjournalists should get in touch with the Freedom of Information Clearinghouse, P.O. Box 19367, Washington D.C., 20036.

Many states also have Freedom of Information laws, known as "sunshine laws." Usually these laws (which apply to city and county governments as well as the state) have fewer teeth than the federal FOIA, but if you keep pushing, threaten to sue, and gather the support of one or more state legislators or city council members you can generally get at least part of what you need. If your state or city has an ombudsman's office, enlist its help. Note that state and city agencies often keep duplicate files: When reporters in Springfield, Massachusetts, were stonewalled by the Springfield License Commission regarding certain mob-connected liquor licenses, they turned to the state Alcoholic Beverage Control Commission for the duplicates. In general, the tricks for using the federal FOIA will apply to state sunshine laws with only minor variations.

## 2.6 Computers and Databases

Thousands of commercial and government databases can now be accessed by any home computer user. The range and depth of information available online is awesome. Sitting at your terminal, you can access everything from Department of Motor Vehicles auto registration data through synopses of the Ph.D. dissertations of every accredited university in North America. You can search the full text of over 200 daily newspapers, in some cases going back ten years or more, finding every mention of a person's name. You can call up on your computer screen the Securities and Exchange Commission filings of every publicly held U.S. corporation. In some localities, you can search state court plaintiff/defendant indexes and county tax assessment records.

To conduct online searches, you need an IBM-compatible computer, communications software, a modem (the device that connects your computer via telephone to the database vendor's computer), a printer, an instruction book, access to at least one database network, and a few evenings' practice. As of 1991, all the equipment could be purchased for under $1,000 with careful shopping. For tips as to computer brands, see the various computer magazines and users' groups.

Generally, an end user of databases does not purchase access directly from the database producer. Instead, the producer licenses a vendor to market the database. The vendor makes the venture profitable by offering a large number of databases from different producers within a single database system. Each database is downloaded into this system, which includes the vendor's value-added software so that all databases on the system can be searched in a uniform and "user-friendly" manner.

One of the leading vendors is Dialog Information Services, Inc. Its DIALOG Information Retrieval Service offers over 400 academic, business, and newspaper/periodicals databases containing over 200 million records. Another large vendor is Mead Data Central, which offers LEXIS, a vast database system for lawyers, and NEXIS, a system of newspaper, periodical, and business-oriented databases.

The main problem with online database searches is the cost, which can run you as much as $200 per hour of connect time (as opposed to no money—but often a much greater expenditure of time—in using reference books at the public library). One way to contain costs is to go online at night on special rate plans such as Knowledge Index, a DIALOG service providing access to over eighty databases from 6 PM to 5 AM.

Another problem is the start-up fee and/or the annual or monthly subscription or minimum usage fee charged by vendors. These fees are not a burden if you are subscribing to only, say, DIALOG ($42 initiation fee with $35 annually thereafter). The fees can get out of hand, however, if you sign up with several of the major database vendors at once, and if you need access to investigative databases from specialty vendors.

Many database searchers end up using a gateway service to access multiple database systems. Essentially a gateway service is a "vendor of vendors"—it gets you into the various database systems via a single subscription fee. Other advantages: you use a single set of search routines, pay only a single monthly bill to the gateway, and minimize costs by formulating search questions offline. CompuServe's IQuest is among the most popular gateways, offering access in whole or in part to major database systems such as DIALOG, ORBIT, WILSONLINE, NewsNet, VU/TEXT, and DataTimes. (The latter two vendors offer full-text searching of scores of newspapers—a resource we will emphasize in later chapters.)

An alternate mode of accessing databases is via CD-ROM (compact disk with read-only memory). Here a customer purchases or leases the database and periodic updates for direct use with special equipment—a

CD-ROM player with controller card and special software. Although the customer avoids paying for online connect time, the one-time-only cost of purchasing CD-ROM databases is usually too much for an individual researcher. Research libraries, however, are a major customer for CD-ROM databases ranging from magazine indexes through regional telephone white pages. Thus, CD-ROM brings full-text searches within the reach of many people who cannot afford online database searching—they can do their searching for free at the public library.

Most local, state, or federal government databases (e.g., Federal Election Commission databases) are on magnetic tape. These can be purchased or leased at a relatively modest cost from the particular government agency or its designated vendor. To use magnetic tape databases you need a 9-track tape drive with controller card, and the software to download into your personal computer. The Missouri Institute for Computer-Assisted Reporting (MICAR) at the University of Missouri can give you advice on government magnetic-tape databases (see its monthly newsletter, *Uplink*). If you purchase or lease a magnetic tape database, you can send it to MICAR, they will mount it in their computer, and you can then work with it online from your own PC.

For most journalists (except those involved in detailed analyses of large quantities of government statistics), it's usually more practical to access government-produced public records databases (especially those on the state and county level) via an information broker. The latter can link you for a modest fee to a large network of vendors in every part of the country who have each purchased one or more of these databases and added their own software.

To keep informed of what's available in the database world, see *Computer-Readable Databases*, a softcover reference from Gale that is also available online via DIALOG. It describes over 5,000 publicly available databases, including not just online but also CD-ROM, magnetic tape, and diskette. Another major directory, *Cuadra Directory of Online Databases*, is available online from CompuServe as well as in hardcopy. Also essential is *CD-ROMs in Print*, which describes about 2,500 CD-ROM titles and is indexed by subject.

Information USA's *Federal Data Base Finder* describes 4,000 free and fee-based federal government databases and specifies the modes in which they are available to the public, e.g., magnetic tape purchase or lease, online access direct from government, and online access via commercial vendor. The directory also specifies those government database producers who will perform searches for people who don't have a computer. The same publisher also offers the *State Data and Database Finder*, which covers all fifty state governments.

Profiles of almost 4,600 database producers, vendors, and gateway services are found in Gale's *Information Industry Directory* (formerly *Encyclopedia of Information Systems and Services*). You should get on the mailing list of the most important vendors and gateways and receive their annual directories and supplements as well as news releases describing their latest products.

## Nonprofit Database Search Services

If you feel a complicated search is going to cost too much, either because you lack the skill to do it quickly or because the particular database is just too expensive to use (e.g., NEXIS or LEXIS), you can always go to your central public library or a local university library.

First, check if the library has the given database on CD-ROM; if so, you can search (and cross-search) for free at your leisure following easy instructions. Note that CD-ROM publishing is beginning to move into the mainstream: There are obvious advantages to a system that can store the equivalent of 250,000 pages on one compact disk for full-text searching without any online expenses.

If the database in question is not on CD-ROM, go to the library's online search facility. There, your search will be conducted at below commercial rate or even at cost by a highly skilled professional.

To find such help, look in Gale's *Online Database Search Services Directory*, a comprehensive guide to both nonprofit and commercial search services. This book indexes search facilities by online system accessed, databases searched, subject areas searched, and geographic area. Let's say you are in Cleveland and you need a database search of newspapers nationwide. In *Online Database Search Services Directory* you find that the Cleveland Public Library accesses NEXIS and DIALOG (both of which offer full-text searches of many daily newspapers). You learn that the library's search service is available to anyone and that it is either without charge or on a fee basis, depending on the complexity of the search. You also learn that the library requires that you be interviewed in person or over the phone to prepare for the search, and that you sign a search agreement. This is a typical example; similar arrangements can be made with public and university libraries from Idaho to Florida.

Journalists who need help with a database search can contact Investigative Reporters and Editors (IRE) at the University of Missouri School of Journalism. IRE will arrange for a reference librarian at the university library to perform your search at cost, if you are a member of IRE. Note that this professional organization is open to freelancers as well as staff reporters.

## Specialized Databases For Investigators

Certain types of databases are of special interest to investigators. These include:

- Mailing list companies' residential databases—compiled from telephone books, crisscross directories, U.S. Postal Service change of address files, magazine subscription lists, and other sources. The largest have upwards of a hundred million names.

- Credit-reporting databases—for instance, TRW's Credit Data, which contains information on 133 million people. Credit-reporting agencies are barred from giving out credit information on individuals except according to strict privacy guidelines, but this does not apply to non-credit "header" information on credit reports (address, date of birth, social security number, spouse's name, etc.). The major credit bureau databases can be searched for noncredit data via an information broker (see below).

- Local and state government public records databases. Many of these records, once kept in ledger books or card files and/or on microform, are today fully computerized and are available to personal computer users either directly online (or downloaded from magnetic tape) or via a vendor. In some cases, private firms are compiling databases of their own from public records—they actually have employees down at the courthouse typing everything into laptop computers. As noted above, these databases have become linked into networks through which you gain access to all through a single gateway.

The key to entering this new Sam Spade cyberspace are the information brokers (a list is provided in the Appendix). They will provide you the gateway to the various investigative databases so you can do your own searching. Or you can let the broker do the search for you and send you the results online or by phone, fax, or E-mail. The brokers often simply subcontract the search to the primary vendor of the database in question. If there is no database in a particular state or county for the records being searched, the broker will arrange through the online investigative net for a hands-on search by a documents retrieval service in the city where the records are kept.

The *Online Database Search Services Directory* and the *Encyclopedia of Information Systems and Services* list information brokers in every part of the country. One firm that specializes in helping newspaper reporters as well as private investigators is Datafax in Austin, Texas (see Appendix). It is operated by Ralph D. Thomas, the author of *How to Investigate by Computer: 1990*. Datafax requires a $60 user deposit but no monthly minimum charges. It will search driver's license records throughout the country, conduct a sweep of major credit agency records to find someone's social security number (SSN), run a bank account search ("input name and address, and we can locate for you the bank and account number of the party in question"), and match up a vehicle license plate number with the name and address of the vehicle's owner in any state. The cost for an SSN trace is $16 for each credit bureau network which is searched, and $38 for three networks.

A word of caution on databases: They are not a form of magic. Many investigative journalists will tell you that database searches rarely produce more than a tiny fraction of the information they need in preparing a thorough background report or investigative news series (of course, the databases do produce a large number of important leads to be followed up). In addition, databases are only as accurate as the people who do the information gathering and inputting. You are thus advised, if a particular piece of evidence obtained online is crucial to your investigation, to get a certified copy of the actual document from the official record-keeping agency.

For further information on investigating by computer, Ralph Thomas' manual is an excellent introduction (see bibliography).

# 3.

# Some Basic Techniques and Procedures

## 3.1 Getting Facts Fast

Many times in the course of an investigation, you will need a fact immediately: the home address of a corporate executive; the law school background and year of graduation of a local attorney; the name and address of a trade association; a list of a major corporation's subsidiaries.

One way to get these facts is by online computer databases, but if you use them to answer every minor question your costs will soar. Fortunately, there is another way to check out vexing details without leaving your office. The central public library in many cities operates a telephone reference service staffed by professional librarians who have at their finger tips hundreds of basic reference works.

This service is free and only takes a few minutes—but keep your requests simple. If the librarian doesn't have on hand the book with the answer, her or she can at least tell you its title and refer you to the reference service at another library that does have it.

Library reference services can best help you if you know exactly what you want. This is why you should have a copy of *Directories in Print*. Find there the book you want searched before you call the reference service.

I have found that the public library reference services vary in quality. In some cities, the line may be perpetually busy or the number of directories at hand relatively small. In other cities, the service is excellent. If you can't get through to your local reference number, simply call a library in another city (the *National Directory of Addresses and Telephone Numbers* gives the central number for hundreds of public libraries). In researching an article on the Teamsters Union a few years ago that had to be finished fast, I used the telephone reference services of fifteen libraries around the country during a single week.

The Library of Congress' telephone reference number is (202) 707-5522. Only call it if local public libraries can't help you.

## 3.2 Telephone Information "Pyramiding"

Public library reference librarians generally will allot only a few minutes to any given caller. If you have a question that is too complicated for them, the federal, state, and local governments maintain a vast cohort of public information officers, press secretaries, and legislative committee staffers who routinely answer questions from the public or refer the callers to the appropriate government expert. To find out who to call, look in the guide books described in 2.1 (Lesko's *Info-Power* is especially strong on government sources of free public information). If you're stumped, call the legislative office (*not* the community office) of your elected representative on the relevant level of government for advice. Your regional Federal Information Center may also help you identify sources of information at federal offices either within the region or in Washington. Once you reach the best expert in a given agency, he or she may steer you to an official in another agency or someone in the private sector for additional information.

You can also seek information on your own from a wide variety of private nonprofit organizations, ranging from trade associations through think-tanks. Get a telephone reference librarian to look up likely prospects for you in the subject listings of the *Encyclopedia of Associations.*

Other good sources include newsletter editors (your local telephone reference librarian can find the right newsletter by looking in *Newsletters in Print,* which is divided into thirty-three subject categories) and corporate public relations or communications directors (their names and telephone numbers at any large corporation can be found easily in any of several business directories). I received much help from both newsletter editors and corporate PR types in researching this book.

With the vast range of sources described above it is possible to find an answer over the phone quickly and efficiently, even to quite arcane questions. You are passed on from general to specific experts, or from experts in one aspect of your question to experts in another aspect. I call this "information pyramiding." Once you get the hang of it, it rarely take you more than three or four calls to zero in on the person with the most definitive answer.

## 3.3 Collecting and Filing Your Documentation

You can master every technique for gaining background information, but it won't do you much good if everything gets lost in a huge pile of unsorted documents on your desk. This doesn't matter much if you're doing a relatively simple investigation, but if you are preparing a major piece of investigative journalism or compiling a deep background report that involves parallel backgrounding, you will find yourself accumulating documents and also notes from telephone inquiries and interviews at a rapid pace. It is essential that you devote a period of time at the end of each day to filing and crossfiling this material. If you let this go for more than a couple of days it will get out of hand. Not only will you not be able to find things, you will find it very difficult to plan out your future research efficiently. You may discover an important lead one day, set it aside thinking you'll follow it up next week, and then completely forget it.

There are as many filing systems as there are journalists. Here's my system: First, always have on hand several dozen filing folders at the beginning of your investigation. Use legal-length folders—if you use letter size you'll have to fold documents before inserting them, which makes the file harder to search. When you label a folder use a pencil so you can erase and use it again. As the documents accumulate during a backgrounding assignment, make a folder for each aspect or important event or stage of subject's life (e.g., college years, first job, 1987 lawsuit, divorce proceedings, 1989 federal prosecution—you might also want to construct a chronology). Also make a folder for each of subject's associates, business entities, etc., as well as folders on appropriate background topics. Before you file a document, highlight the important parts with a yellow marker. You will find that many documents or interview notes relate to more than one topic; in such cases, make multiple photocopies and place a copy in each relevant folder with highlighting specific to that folder. If you don't have a photocopier in your office, you can place the document (with a red "X" on the front page) in the file to which it has greatest relevance and then insert in the other file(s) a yellow-pad sheet bearing the reminder "See—1987 lawsuit file." You will find that some documents are relevant to almost every folder; in such cases, make an "Urgent—Multiple Use" file, and keep such documents there to be constantly consulted. When a folder begins to fill up, divide it into primary relevance and secondary relevance folders.

A rule of thumb: If you're not spending at least 10 percent of your time doing such filing during a major investigation, there's something wrong.

My filing system flows from my approach to the gathering of documents: Photocopy everything of possible relevance in a court file, newspaper clippings file, library reference volume, etc. This is especially important in the early stages of an investigation: You won't know what's most important until you fit it into the larger picture.

When you find a damning public document—for instance, a deed of sale showing that a local politician bought a parcel of land two weeks before a zoning change raised the land's value dramatically—always get a certified copy from the county or city clerk's office immediately. Such documents have a habit of disappearing from the public files within a few days of your call to the politician to question him or her about it.

Finally, one of this manual's most important pieces of advice (for which you will thank me many times over): *whenever* you go to the courthouse or the public library, carry a mimimum of $20 in rolls of quarters and/

or dimes for the photocopying machines. Public facilities don't always provide change, and when they do, you have to stand in line. And don't depend on getting your rolls of change at a bank around the corner; some branches nowadays will only provide them for depositors, and the banks usually close at least two hours before the courthouse file room.

## 3.4 Eliciting Information From Sources

Skip tracers and private investigators use various ruses to gather information. For instance, they may call up a friend of the skipped debtor and say they represent the estate of a relative who has left the skip some money. The friend, wanting to be helpful to his or her pal, usually will reveal the new address or phone number.

Many times ruses aren't intended to trick people into giving out closely held information, but simply to make them feel comfortable about discussing something they have no strong reason *not* to discuss. The ruse allays their natural suspicions of a stranger asking questions about a third party. It also gets around their natural antipathy towards bill collectors.

Apart from ethical concerns, the use of a ruse makes it more difficult to go back to the same source later. The person you manipulated into revealing an unlisted telephone number may turn out later to be the key to far more important information. But the trick you used the first time you contacted them may render subsequent questioning somewhat embarrassing.

Journalists can avoid ruses under most circumstances, because they have a built-in reason for their snooping—they are working on a news story, and their questions are part of that story. Most people understand this and respond politely and without suspicion. If they know the journalist is on a deadline, they'll often drop a pressing task to answer his or her questions on the spot.

Another advantage journalists have is that their craft emphasizes tracking down opponents and enemies of their subject. Whereas a skip tracer might pump a friend of subject for information using a ruse, the journalist will naturally gravitate to subject's enemy and get the same information (and more) in an above-board fashion.

Whether you are a journalist or a skip tracer, your success in eliciting information depends on your mind-set. Remember always that the average person has a natural desire to be helpful if they're not too harried at the moment or feeling sick. Approach each contact as if you were lost on a country road and seeking directions to the nearest town. As the conversation develops, let them know that you are trying to do a job on a deadline with your editor pressuring you (anyone who works for a boss can identify with this). Behave as if there's no question in your mind that they will help you. The more you believe this, the more it will come true unless you sabotage things by being rude, unctuous, overly aggressive, etc. If you haven't learned basic human communication skills, there are many good books and seminars on the subject.

If you need a source's help on anything more than a simple question or two, you should strive to make your investigation personally meaningful to him or her. This means stimulating the source's sympathy, curiosity, or self-interest, and appealing (when appropriate) to his or her personal convictions. For instance, if you are working on an exciting or colorful story, try to get the person interested in what you've found out. If you are working on a story about a crooked politician, appeal to the person's indignation and desire to do right. If you are working for a defense attorney, appeal to the potential witness's natural sympathy for a little guy caught in the toils of the legal system.

If a contact refuses to talk and/or slams the phone down, try them again in a few days with a slightly different pitch—they may just have been in a bad mood, or you may have used the wrong approach the first time. I've had people scream and curse at me on the phone one day, and be perfect lambs when I called back twenty-four hours later. Also, when working on long-range projects, I've had sources adamantly refuse to talk at the outset but be eager to talk six months or a year later.

Likewise, if a contact gives you only very limited and vaguely worded information (but you know they know much more than they're saying), don't give up. Thank them politely and ask if you can call them again when and if your research generates further questions. Most people will say yes, if for no other reason than just to get you off the phone or out of their office. But it lays a basis for calling them back without seeming too pushy. Again and again, I've found that cautious bureaucrats or frightened cult defectors will open up marvelously during a second or third conversation if you don't push too hard.

If you are investigating the sinister Mister X, don't forget the most important source of all: Mister X himself. A journalist must get his subject's side of the story at some point, but when to do so is a complicated question. On the one hand, interviewing your subject early in the investigation may save you a lot of unnecessary digging: First, you may find Mister X is not quite the villain you thought he was. Second, you may find him willing to turn over important information in order to placate you, justify himself, or shift the blame onto one of his associates. Third, Mister X may not perceive his behavior the way you do; what is sinister to you may seem admirable to him and not at all to be covered up. For instance, I once spent weeks digging into a Teamster official's relationship to a far-right organization before calling him. I might as well have called him the first day and saved myself all that digging: He was proud of the connection and discussed it freely.

On the other hand, tipping off your subject too soon may trigger action to close off important avenues of information. In investigating an elected public official, I suggest that you contact him or her early in the investigation with relatively noncontroversial but necessary background questions, then come back later with your zingers. In general, the less controversial or potentially damaging your investigation of a person is likely to be, the earlier you should approach him or her.

Finally, there is the problem of the putatively eccentric or disreputable source. It would be nice if investigative journalism involved only interviewing priests, rabbis, and an occasional professor of business ethics at a local college. Unfortunately, in many investigations the only people with inside knowledge are those regarded as either weird or sleazy—you have to take what you can get. In investigating far-right politics I have had to deal with many such people. Although some journalists have been burned by such relationships, my experience with borderline sources is that they often provide information as reliable as—and certainly a lot more interesting than—the information provided by impeccably respectable people (who, by the way, have their own methods for covering up the facts).

The basic rule in dealing with so-called eccentric sources is to afford them the same respect as anyone else, and never violate their trust on grounds that they somehow don't deserve the same straight-dealing as ordinary folks. This does not mean becoming their doormat. You may find that they have a narcissistic tendency to call you at 2 AM collect from Alaska with their latest brainstorm. An experienced journalist-therapist, however, knows how to put an end to these excesses with a bit of Pavlovian conditioning (if they call after midnight, hang up and then be noticeably cool to them at the outset of the next conversation).

## 3.5 Interviewing

An interview, as opposed to a brief telephone conversation with a contact, involves formal questioning based on mutually agreed ground rules. The first thing to remember about interviewing is: Do your homework. If your subject is an expert in a field and you come to the interview totally ignorant, he or she will be annoyed and thus be much less inclined to cooperate. Being prepared also means drawing up a list of the questions you want answered and putting them in some order of priority. You should ask yourself what the person might know *beyond* the obvious. Time and again I've interviewed people and failed to probe far enough, failed to get to the key question—and it was usually because I called them hastily without preparing beforehand. (If, after you hang up, you realize you've missed a key question, call the source back immediately—this may actually start a longer and far more illuminating conversation.)

Whether you interview a person on the telephone or in person, you have a choice of taping the conversation, taking notes, or both. Taping someone on the phone without their knowledge is illegal in several states (and since you can't use the illegal tape in court or to convince your editor, why do it?). However, if you ask a person's permission to tape them over the phone, you may screw up the interview. If they consent to be taped, they often will be excessively cautious in what they say (remember, they are talking to a stranger they cannot see). If they don't consent to be taped, they'll think you have the machine on anyway, and thus will talk much less freely than if you'd never raised the idea. These problems will probably not arise in a phone interview with an expert providing noncontroversial background information; even then, make sure he or she feels comfortable about it.

When you interview someone in person, there is more leeway for using a tape recorder. If you have established rapport with them and don't intend to ask anything that might compromise them (and if the conversation is *on* the record), I would say the tape recorder is appropriate—just don't place it in their direct line of vision while they are talking. Otherwise, simply take notes.

Even when you use a tape recorder, taking notes will save you from having to listen to and transcribe material later that may only be of minor importance. Also, taking notes insures against such common occurrences as the recorder not working properly or background noise ruining parts of the recording.

Both cassettes and notes should be marked for identification immediately: Put on the cassette label the name of the person interviewed and the date of the interview. On the first page of your notes, put the name, date, time, and either the place at which the interview was conducted or the telephone number that you called subject at (and the number from which you called). Take notes on the front of each page only; number each page afterwards (1 of 3, 2 of 3, etc.) and print "END OF INTERVIEW" right beneath the last line of notes. While the interview is still fresh in your mind type up the notes on your word processor, filling in the gaps from memory but using quote marks to distinguish subject's actual words from your memory additions.

When this is done, do not destroy the original handwritten notes (if you are a journalist you will need them in dealing with editors and your newspaper's libel attorneys before publication, and possibly in a libel defense thereafter).

Note that with any interview you should get the correct spelling of the person's first and last name, the middle initial, and any generational designation. In interviewing public officials or corporate officers be sure you get their correct title AND the correct full name of their agency or company. There is nothing more annoying than to read a news article (or a private investigator's report) in which the chairman of the city board of education is referred to as the "director" of the "division" of education. In dealing with a high-level official, get this information from his or her private secretary or an aide rather than bothering him or her with it during the interview.

If you interview someone in his or her office, observe carefully your surroundings: a diploma on the wall or pictures of family members on the desk may provide important leads. Also note the names of the interviewee's private secretary, the receptionist, etc. These people may hate their boss and become important sources later on.

Journalists make a variety of agreements with sources about how they will use the information. Ideally the entire interview should be "on the record" (i.e., you can quote the person by name on anything they say). If they decline to be quoted, however, you can put forward various alternatives. First, suggest different terms for different aspects of the interview, i.e., some of it on the record, some of it not. (Often, to make your story, you will only need a single quote on the record.) If your source still refuses to be named, ask if you can quote him or her on a "not for attribution" basis, e.g., as "a former top aide of Senator Foghorn" (no name given). Only if this doesn't work will you offer to do the interview as "background" (you attribute it only to vague "official sources" or "well-informed sources") or "off the record" (you only use the information if you can verify it from other sources or from documents and never refer even in the vaguest terms at all to the existence of your original source).

In discussing attribution with government officials or prosecutors, another tactic to overcome their skittishness is to offer (again, as a fall-back position only) to accept the information as attributed to them by name, but not for direct quotation (you paraphrase them; if they get in trouble about it, they can always say you misquoted them). A further fallback is to offer to accept the information for attribution to an anonymous "spokesman" of the source's agency or organization. Obviously you would in most cases only raise these arcane distinctions with someone who is sophisticated in dealing with the press. With ordinary people, I would stick usually to three categories: on the record, attribution to a clearly defined but unnamed source, and background (attribution only to a vaguely defined unnamed source, e.g., "a former cult member").

In preparing for any interview, the question arises: Should you be the first to raise the question of attribution? In dealing with most government officials (except "whistle-blowers") or most politicians, I would say: Let sleeping dogs lie, and just assume everything is on the record. If the official is stupid enough not to set ground rules, he or she must pay the price. But in dealing with ordinary people, it's not so simple. Often they will tell you something openly, and then say halfway through the conversation or at the end, "don't quote me on this." If you quote them anyway, they'll feel aggrieved and won't cooperate in the future. It is thus usually best to establish ground rules for the interview if you anticipate further dealings with the interviewee. However, I have sometimes had people agree to go on the record, and then change their mind and call me at home afterwards, begging me not to use their name. If I really believe that printing their name will create hardships for them, I'll usually accede to their request.

An important distinction to keep in mind in your interviewing is that between an anonymous source and a confidential source. The former is a source whose identity you will not reveal in your article or in conversations with third parties but whose identity you might disclose in a libel suit or other special circumstance. A confidential source is one whom you have agreed to keep secret at all costs, even if you are jailed for contempt of court as a result. This should be agreed on at the outset if the source is giving you sensitive information. Obviously, you would prefer the source to be anonymous rather than confidential. I have had sources who began as strictly confidential agree to go public later on, as we got to know each other better.

Special problems arise when a journalistic source (say, a government "whistleblower") has extensive inside information regarding a major story. A single interview often will only scratch the surface of what they know. Even a half dozen interviews may fail to uncover a key piece of information which they have but which they don't know is important and which you don't know enough to ask about. As you follow up leads from the earlier interviews and get a clearer picture, you will have to come back to these sources again and again. All the more reason to be absolutely honest in your dealings with them: Maintain confidences and be careful not to misrepresent or misquote what they tell you. If you disagree with their interpretation of something, let them know forthrightly—don't spring it as a surprise in one of your articles. (I often will read the complete draft of my story to an important source; not only does it increase his or her sense of commitment but it almost invariably uncovers subtle errors of emphasis if not outright errors of fact.)

Beyond all this, there is an art to interviewing—of how to do it so as to extract the maximum information. This topic is sufficiently complicated that entire volumes have been written on it (see bibliography).

## 3.6 Getting a Signed Statement

If you are working for an attorney, or if you are a journalist working on especially sensitive material, you will want to get signed declarations from some of the sources you interview. Some interviewers may do this with a laptop computer and portable printer. The method described here involves the good old yellow pad with ruled lines and a pen (never a pencil). Write at the top of the first page "Declaration of John R. Doe." Begin the text, "I, John R. Doe, was interviewed on [date] between [starting time] and [ending time] at [place of interview] by [name of interviewer] who is a reporter for [name of publication] . . ." (or: ". . . who is an investigator for [name of attorney] . . ."). Then write down the source's occupation, place of employment, and home address, followed by each relevant fact of which the source has direct personal knowledge; do not include any mere speculations, no matter how juicy. Read each sentence to the source before you write it down, so he or she can affirm the accuracy. Keep the statement as brief as possible, locking the source into an affirmation of the absolutely crucial facts only. If you need more than one page, number them "1 of 3," "2 of 3," etc. When finished, ask the source to read over the entire declaration carefully. If the source wishes to make changes at this point, he or she should initial each change.

When all changes have been made, the source should sign each page and then write in his or her own hand at the bottom of the last page that he or she has read the above X pages and X lines (count the lines on the last page) and knows the contents to be true. Finally, have the source sign this affirmation on the spot or take them to a notary, who will verify their identity, witness the signing, and stamp the signature page with a notary seal. If you do much interviewing for attorneys, you should become a notary yourself. If you're a journalist who happens to be a notary, do *not* notarize the statement yourself. If you're sued for libel, a statement notarized by yourself is not going to be as credible.

In most instances, a simple declaration (signed but not notarized) is sufficient. The notarized statement—or affidavit—is chiefly useful when the source is telling you explosive and potentially libelous things. It's a way of putting the source to the test—if they are lying or wildly exaggerating, they will most likely back off from giving an affidavit. Even when you fully trust the information, an affidavit may be useful because it takes the source's commitment to a higher level—they become less likely to back out later. In addition, affidavits are sometimes useful when a source insists on talking not for attribution or on background only; having the statement to show to your newspaper's editor and attorneys will help to convince them the information is reliable in spite of the source's insistence on anonymity.

Note: whenever you get a signed statement, also get from the source the name, address, and telephone number of at least one person who will know how to contact them if they move.

## 3.7 "Advertising" for Information

When a person is searching for a missing relative and the trail runs dry, he or she may try placing classified ads in selected publications, asking those with information to come forward. Families searching for an abducted child may pass out flyers which include the child's picture and a reward offer, or they may go to the media. The police often circulate wanted posters and seek the aid of the TV program "America's Most Wanted." Well, journalists can do this also: Jessica Mitford, in preparing an expose of the Famous Writers School, placed a classified ad in *Saturday Review* (a magazine whose readers she felt were likely to have fallen victim to the correspondence school's scam) asking former students to contact her.

Journalists (or PIs who are part-time journalists or have a journalist friend to feed a story to) can do a more subtle but often more effective type of "advertising." It's called going with what you've got. You take the partial, imperfect story you've uncovered and publish it as a means of attracting sources who can tell you the rest of the story.

I first learned about this tactic while working for the Manhattan weekly *Our Town*. Assigned to a story on local newpaper distribution companies and "swag" (stolen newspapers), I was able to write a story describing how the scam worked but with no hard facts on who was behind it. *Our Town* published the story, and within hours we were getting calls from anonymous persons in daily newspaper circulation departments—and also newsstand operators—offering us information about the role of a corrupt local union. A similar thing happened several months later when we did an article on right-winger Lyndon LaRouche: Within days, calls were coming in from defectors from LaRouche's organization who had a wealth of information. Some of it was about LaRouche's hidden control of one of Manhattan's largest computer software companies. We published a piece on this, eliciting, again, a spate of calls, this time from former clients and employees of the firm in question. One morning I came into the office and found on my desk a complete computer printout of the firm's general ledger, apparently dropped off by a disgruntled programmer.

This technique worked in the above cases because the published articles were so easily available to people with the information we needed. The "swag" scam was centered in Manhattan, the LaRouche organization had its headquarters in Manhattan, and LaRouche's computer company did much of its business in Manhattan. People with knowledge of these topics could pick up *Our Town* for free in midtown banks and other busy locations, and all three articles were on the front page. The lesson is: Use your news articles as the journalistic equivalent of a wanted poster to find the sources you need. If your article appears in a paper that reaches the right audience (and that audience, depending on the topic, could just as well be a specialized newsletter with only a few hundred subscribers), the results may surprise you.

If you specialize in a particular journalistic topic—or have developed a reputation for going after villainy across the spectrum—you can do advertising on a much broader scale: Go on radio and TV talk shows, solicit speaking engagements, hold press conferences on your latest article, do newspaper and magazine interviews, and get other reporters around the country to quote you as an expert. Over the long run, this can pay big dividends. Much of the information for my 1989 book on LaRouche came from sources who first heard me on talk shows, or saw my name in an article by another journalist, or came up to the podium after one of my speeches.

Another approach is to cultivate people in organizations concerned with your topic who will then refer potential sources to you. In the mid-1980s, I had over a dozen Jewish, civil rights, labor, and anti-cult organizations passing on my name to victims of the LaRouche organization; this also produced major leads. In addition, if you let other journalists around the country know that you are focussing on a given topic, they will call you when they are working on a relevant piece, and, in return for your help, share their findings with you and introduce you to their sources.

I believe in getting other journalists involved to the maximum extent. If the story is a big one, there's always room for several people working on different aspects. What one turns up will help the others. The tracking of LaRouche in the early 1980s, for instance, was the work of a network of journalists all over the country. Whenever an article about LaRouche by freelancer Russ Bellant appeared in a Detroit paper, it would stimulate local ex-LaRouchians to contact Bellant who would never have known to contact me in New York or Chip Berlet in Chicago or Joel Bellman in Los Angeles (and vice-versa). Unfortunately reporters on major dailies are often too paranoid or competitive to practice this cooperative approach.

One of the best forms of advertising is simply to have a listed home phone number in the telephone directory. This may produce some annoying messages on your answering machine (and certainly you might want to have your address unlisted) but to have an unlisted home phone number is, for an investigative journalist (especially a freelancer without a fixed office), the cardinal sin in my opinion.

## 3.8 Caller ID

A hard-hitting investigative series often produces threatening or harassing phone calls—or calls from imposters trying to find out how much else you know. I suggest that you subscribe to a Caller ID service if available in your area. This will tell you the number of the phone from which the call is being made, and you can then reverse the number in the crisscross directory to find out the subscriber's name and address. Knowledge of the identities of crank callers is not just good self-protection: It also can help you fill out the list of your subject's known associates. Some of them may later become sources (especially if you are investigating cults—an area in which today's true believer is tomorrow's defector). Of course, the usefulness of Caller ID is limited by the fact that many crank callers operate from pay phones, and also because in some areas the phone company allows callers to block their number by entering a three-digit code before dialing. Still, a person disturbed enough to make crank calls may lack sufficient impulse control to use the blocking code consistently.

Caller ID is also useful if you are dealing with an anonymous source who is not yet ready to tell you his or her identity. Learning the identity surreptitiously and then gathering a little background information will enable you to decide whether or not the person's information should be taken seriously.

# 4·

# Finding "Missing" People

## 4.1 Overview

A major problem in many investigations is locating the seemingly hard-to-find source or witness. Usually such people are not deliberately hiding out (or at least not hiding out from *you*). Many prove willing and even eager to talk when and if you find them at their residence or place of employment. But what if you only have an outdated address for them? Or if all you know is that the person lived somewhere in the Boston area many years ago? Or if his or her phone has been disconnected and there's no forwarding number? Or if the new number is unlisted?

Difficulty in finding people stems partly from the high rate of mobility in the American workforce. A 1984 University of Michigan study found that 30 percent of interviewees nationwide had lived at their most recent address for less than two years. An earlier survey estimated that 16.6% of all citizens of voting age had changed their place of residence within the previous year. The close-knit neighborhood in which folks sit on the front porch in the evening and everyone knows everyone else is largely a nostalgic memory.

This trend is paralleled by a growing desire for privacy, even a kind of secretiveness. During the 1980s, the percentage of telephone subscribers opting for unlisted phone numbers (numbers unpublished in the telephone directory and unavailable through directory assistance) rose dramatically. By 1988, one out of four Americans had an unlisted number; in California, the rate was 38 percent unlisted. In addition, the percentage choosing to list their name and telephone number without their address (or with only a partial address or a P.O. box) increased sharply.

Such problems are offset by a variety of information systems, both online and manual. Over the past decade the time-honored methods of the skip tracer based on marketing directories and pretext phone calls have been supplemented by powerful new database search techniques. Today the majority of Americans are listed in the giant databases compiled by credit-reporting agencies, mailing list vendors, state motor vehicle departments, etc. Also, crisscross directories can be used far more effectively today in a computerized format.

But databases are only as good as the information fed into them, and this information is often inaccurate or out-of-date. Many people contrive to drop out of the electronic information net, or to build a parallel database trail under a different name and at a different address. Their motives run the gamut: an ex-husband who wishes to evade child-support or alimony payments, the criminal suspect who has jumped bail, a deadbeat who owes thousands of dollars to department stores, a cult member who believes her parents are instruments of Satan plotting to have her tortured by deprogrammers.

People who don't want to be found use a variety of predictable tactics. For instance, they put their telephone or utilities in someone else's name, or they use a false Social Security Number (SSN) on job applications. But most such people are inconsistent in using these tactics. They may give a phony SSN at one job but forget and give their real SSN at the next. They may list their phone number in their girl friend's name, but list a utility bill in their own name. Finding them just means searching systematically through the various records systems until you find the weak spot.

The problem is more difficult when your subject knows how to conceal his or her location and identity in a systematic manner and has the discipline to stick to it. Some of these people go to elaborate lengths to construct a false identity, even obtaining a birth certificate and other ID in the name of someone born about the same time as themselves who died in childhood (this is called "paper tripping"). If the FBI can't find these people, you may not either. But at least you can give it a good try before cutting your losses.

## 4.2 Short-cuts to Finding People

Many of the techniques in this chapter presume that you are starting cold—that you know nothing about your subject except his or her name. In most cases, however, you are *not* starting cold—you already have a smattering of information to guide your search. Begin by listing each fact or rumor you've heard, along with the name of the person from whom you heard it. Get back in touch with this person and ask for further details, no matter how "insignificant," about subject's family background, associates, hobbies, etc. (You might want to compile a model questionnaire for this purpose.)

In listing what you already know, and then looking over the various search options in this chapter, you may find that your problem is solved almost before you begin. For instance, is subject a licensed or credentialed professional? Stop right here. The addresses of professionals—doctors, nurses, chiropractors, lawyers, college professors, engineers, etc.—are almost always available in nationwide or local professional directories at the public library, or else via the state licensing board for the given profession. Is subject a corporate executive? If so, he or she can usually be found through trade directories or database searches of business periodicals. (If your subject is in one of these two categories, skip ahead to section 4.15 of this chapter.)

If your subject has no reason to evade you, and you are in no hurry to find him or her—and if your reason for the search is a humanitarian one (e.g., locating a lost parent)—turn to 4.28, which describes various government locator services. Let the Social Security Administration, the IRS, or the Veterans Administration locate your subject and forward a message asking him or her to contact you.

At any time during your search, you can go to an information broker (see 2.6) for a national sweep of databases. The broker might patch you into the National Credit Information Network (NCI), which has access to all major credit bureau databases as well as national telephone/address criss-cross and postal change of address databases. You enter subject's name and SSN or last known address, and are able to search the non-credit ("header") information in credit-reporting databases covering tens of millions of people. News reporters can access NCI through Datafax Information Services (see Appendix); an average "run" costs between $3.50 and $17.00.

In many cases, such a search will elicit your subject's current address or at least a fairly recent one from which you can work forward using the various search techniques described below. However, before calling an information broker I would try the phone books and Directory Assistance—many seemingly hard-to-find people are really only a couple of phone calls away.

In using the techniques that follow in difficult cases, you should proceed systematically, overlooking none of the steps I describe—and none of the peculiarities of each records system. For instance, if you look in backfile parking violations indexes in the New York County courthouse for a subject's address, it may be there—but you may fail to find it if you don't look in the out-of-state indexes or if you fail to note that the New Jersey books sometimes list people alphabetically by first name rather than last name.

Before using any of the more elaborate techniques that follow, study the given section carefully. The most important thing is not the techniques themselves but the underlying principle—that an almost infinite number of angles can be played to get the information you need. Think creatively! Invent your own variations on these techniques as you go along.

## 4.3  If You Think Your Subject is Still in Town

The following techniques may seem very simple, but you'd be surprised how easy it is to foul up a search of the telephone book or a Directory Assistance request.

## Directory Assistance

Many people simply don't realize the differences between this and the phone book. First and most important, Directory Assistance listings are updated daily rather than once a year. (In New York City alone, there are an average of over 20,000 additions and deletions each day.) Second, Directory Assistance uses the SOUNDEX system, which sorts words both phonetically and alphabetically, thus enabling the operator to find all sound and spelling variations of a name. Third, you can find out from Directory Assistance if subject has an unlisted number—confirmation that he or she still lives in the area.

When you use Directory Assistance:

- Always ask the operator to check under subject's name in the business as well as residential listings. Millions of Americans, especially those in the professions, have business listings under their own names.

- If subject is not listed with Directory Assistance, get the listings for anyone else with his or her surname. The phone might be listed in the name of another member of the household, or one of the listees might be a relative or ex-spouse who will know how to reach subject. (This is only practical with Directory Assistance if the surname is not a common one.)

- If there is no listing under subject's first name, get the operator to check subject's middle name or nickname. (If you are looking for me, and you ask the operator for the number of William D. King, you won't find it; if you ask for Dennis King, you will.)

- If a spouse's maiden name is known, have this checked also.

- In many localities, Directory Assistance will search the entire area code at once if you request it. Check all area codes in your greater metropolitan area, including "exurbia." Remember that nowadays people are commuting longer and longer distances. If a person works in New York City, they may live as far away as southern New Jersey. Some government employees in Washington, D.C., commute from West Virginia.

- In most areas, prerecorded voices are now used to give you the number. This means that any special search request, and any request for the address as well as the phone number, should be made *before* you give the operator the name—otherwise, the operator may not wait to listen. NOTE: When searching via Directory Assistance, always specify that you need the address as well as the phone number.

## How To Use A Telephone Book

You might want to begin your search by looking in the collection of area telephone books in your office rather than by calling Directory Assistance (especially if the name is a common one). And if Directory Assistance fails to come up with subject's number you will surely want to do some searching of the phone books. The following points should be kept in mind:

1. The majority of telephone subscribers have listed numbers. Never assume a number is unlisted, even if you think the person has some reason to need an unlisted number. On more than one occasion, I have been contacted by ace investigative reporters who complained about how many calls they had to make to get my number. When I asked them why they didn't just look in the Manhattan phone book (in which I am one of only two Dennis Kings listed) they told me that they just "assumed" I would have an unlisted number because I write about cults.

2. Never assume you have the correct spelling of a name; always check variant spellings. At the beginning of a list of same-spelling surnames, the phone directory often will provide alternate "see also" spellings. For instance, in the Manhattan directory, we look for Larry Kahn. At the beginning of the "Kahn" listings we find the notice, "KAHN SEE ALSO CAHN, CONN." Such notices, however, aren't always included or may not list all variants (the *New Dictionary of American Family Names* is useful here).

3. Look in *all* phone books for your metropolitan area. More than once, I've failed to find someone because I only looked in the Manhattan, Bronx, and Brooklyn directories—and was too lazy to reach on the higher shelf for the Queens and Nassau/Suffolk County directories. Never assume that subject is still living in the particular neighborhood he was in last year. Check the entire greater metropolitan area.

4. Check the complete listings for subject's surname: You may find that subject is listed under his or her middle name rather than first name (or vice versa if you happen to be searching for a listing under the middle name); or that subject and spouse are listed together, with subject's first name after the spouse's, e.g., "Smith, Mary and John" (you may miss John if you don't go through the complete list); or that subject's first name (or middle name in place of first name) is given only as an initial, e.g., "Smith, S." instead of Smith, Susan."

5. Try calling all the same-surname listings if there aren't too many. One of them may actually be subject listed under a first name, middle name, or nickname you didn't know about. Or the listing may be in the name of subject's spouse, son, daughter, a parent, or other relative (especially if the directory is for subject's home town). Or

the listing may be that of an ex-wife who has retained subject's surname. Also, people unrelated to subject but having the same last name (especially those with the same first name but different middle initial) may have received numerous phone calls in the past for subject and may know how to reach him or her. (People with the same first and last name will also probably recall news articles about subject—their own friends will have brought such articles to their attention, asking if the article was about them.)

6. Look under spouse's maiden name or under linked names, e.g., Mary Brown-Smith, Mary Smith-Brown, John Brown-Smith, John Smith-Brown. Check other persons with the same surname as the wife's maiden name—you might happen on an in-law. (For more tips regarding name variations, see 7.1.)

7. Check for any business listing under subject's name. If it's included in the white pages, it will be directly underneath subject's residential listing unless subject uses a different variation of his name for business purposes. If there is no business listing in the white pages, look in the yellow pages. Note that someone operating a small business may designate it by their surname (e.g., King Enterprises) or may use their full name (e.g., William D. King Enterprises).

### Phone Numbers Listed In Another Name

As we have seen above, subject's phone may be in the wife's maiden name. It may also be in the name of another family member, a live-in lover, or a roommate. This may be simply for convenience—Janet agrees to handle the phone bills while Harold takes care of the utilities. Or it may be that Harold is ducking bill collectors, in which case both phone and utilities may be in Janet's name. It also may be that when Harold and Janet moved from their previous address, Harold owed the phone company hundreds of dollars. Getting the phone in Janet's name is thus a tactic to evade that bill and also avoid paying a deposit on the new phone service.

Sometimes people will obtain the telephone account in their own name but list the number under a fictitious name. This is a way to get, in effect, an unlisted number without paying an additional charge. It also means that a searcher cannot confirm that subject is still living in town.

In such situations, subject can probably be found by going to various public records—such as tax assessment or voter registration lists (see 4.22)—where subject's real name will be listed. If you get the address from one of these lists, you can then go to a crisscross directory and find out the name and telephone number listed for that address. Chances are this is either the fictitious name subject is using or else the spouse's maiden name. Call the answering machine and see what name, if any, is given with the message.

# 4.4 If You Think Subject Has Moved to Another City or Region

### Call the Old Number

Look in a back phone directory and get subject's previous number. There may still be a recording telling you the new area code and number. Or the old number may have been taken over by another member of subject's previous household or a roommate who stayed when subject moved out, and who will know the new number. Or the old number may now be listed in the name of a friend, relative, or sublessee of subject who moved in when subject moved out.

At the least you may get the stranger who was reassigned subject's old number and who may have received other calls from people trying to reach subject. If the new holder of the number is a gossipy type, he or she may have picked up some interesting information.

### Call Directory Assistance

As noted above, the operator in many localities will search the entire area code for you (and the entire state if it's all one area code). You can thus check large parts of a given search region quickly against the most up-to-date database available. Unless you have information to the contrary, begin with the localities closest to subject's last known address: The overwhelming majority of people who move their residence remain within the same metropolitan area or state.

The use of Directory Assistance in such a sweep works best if you are searching for an unusual surname.

### Phone*File

Available via CompuServe, this database enables you to conduct an online nationwide search. Phone*File is compiled from telephone directories and other publicly available sources of information. It covers about 83 million households nationwide. You can search for a person by name with address, by surname with city and/or state, by surname with zip code, or simply by phone number. Phone*File carries a $15 surcharge beyond CompuServe's base connect charge; a single search costs about 50 cents. Try all likely variant listings, e.g., spouse's name (including maiden name), subject's middle name, etc. Also see if you can find listings for any known relatives or business associates of subject.

### Regional Telephone Company Databases

The phone companies have developed regional databases of all listed phone numbers in multistate regions. For instance, NYNEX offers the NYNEX Fast Track Digital Directory, with over 10 million residential and business listings, covering all listed numbers in New York and in every

New England state except Connecticut. Updated monthly, Fast Track is available on CD-ROM.

An online version with residential listings only, the NYNEX Electronic White Pages was announced in June 1991. Similar CD-ROM and online services are available from U.S. West and Southwestern Bell. AT&T may soon offer combined access.

These regional databases are in effect giant crisscross directories. With Fast Track, for instance, you can conduct searches based on telephone number and street address as well as name. The ability to conduct street searches means you can get the names and phone numbers of subject's neighbors (however, unlike with a standard crisscross directory, Fast Track does not tell you the length of residence). If you have only part of a name or street address (as from a torn envelope), you can access all listings that include the given sequence of letters. If you have a street address but don't know the city, you can access all such street addresses in all the cities within the database. These search capabilities will be expanded greatly when Fast Track extends its coverage to back-issue phone directories.

Fast Track can be used for free at your public library. But the library version is only updated twice a year—a serious weakness considering that there are 600,000 changes in listings each month.

### Out-of-Town Phone Directory Collections (Hardcopy and Microform)

Major public libraries have collections of out-of-town directories. Although they are rarely complete or up-to-date, these collections can be useful when you fail to find subject via Directory Assistance. You can search at leisure for same-surname or wife's maiden name listings and perform other searches as with your local phone book. Always focus first (unless you have clues to the contrary) on the same state or region in which subject was last listed.

Searches of out-of-town phone books are rendered much easier by University Microfilms International's Phonefiche. This is a nationwide microfiche collection of 2,600 directories covering 50,000 communities (about 70 percent of the nation). Phonefiche is available at many public libraries, but not all will have the full collection. The microfiche, which includes both yellow pages and white pages, can be bought in selective packages; thus, you may find that your public library only carries the microfiche directories for your own state and for cities nationwide with a population of over 500,000. Phonefiche includes an annual printed cross-reference guide: You can look up any community in the United States, and the guide will tell you which multicommunity phone directory or directories covers that community or any portion thereof. Likewise you can look up any phone directory by title—say, the Bergen County, N.J. directory—and get a list of all communities covered in whole or in part by it.

## 4.5 Searching for Subject Via His or Her Previous Address

A person can't always have been in hiding. If you believe subject previously lived in your city, check the back issues of the local phone directory. Start with the most recent editions and work backwards (or with a long-ago address and work forward). Remember to follow the procedures used with the current phone book (wife's maiden name, parents' names, etc.). If subject appears in a recent or not-so-recent back issue, call the number and see who answers.

If you need to search backfile directories in another city, use Phonefiche's nationwide backfile of thousands of telephone directories dating back to 1976. Unlike Phonefiche's current directory collection, the backfile is only available at a few libraries so far. Check with the research division of your area's largest public library.

## 4.6 Postal Change of Address Notices

Once you get the old address, the easiest procedure is to check with the post office. Most people file a change of address notice with the post office so that their mail can be forwarded. Even skips (people who have moved out leaving many unpaid bills) often file such notices, blithely unaware that this is public information. Just send your request for the forwarding address together with a money order for one dollar to the postmaster of the station responsible for mail delivery at subject's old address. You can find which post office this is by looking in the zip code directory.

An alternative method is to mail an empty envelope addressed to subject at the old address with the words "Do Not Forward—Address Correction Requested" on the front of the envelope.

Although these two methods sound simple, there are problems:

- The address you receive may not actually be subject's new address. It may be the address of a close-mouthed friend or relative, or it may be a mail forwarding service.

- Subject may have filed the change of address notice to fool bill collectors, while continuing to live at the previous address. The mail goes to a mail drop in another state, which forwards it to subject's mother's house.

- Subject might never have filed a change of address form and might still be receiving mail at the old address, even though he or she has moved out. A friend has taken over subject's apartment and is forwarding the mail to subject in a separate envelope (or subject stops by once a week to pick it up).

The local post office usually keeps change of address notices for one year. However, you should still make a request even if a longer period has passed. Subject may have requested that the post office continue forwarding his or her mail for several additional months. Or subject may have waited several months after moving before filing the notice (one can think of many reasons for this, from difficulty in finding a new fixed residence to an extended alcohol or drug binge). Also, the postmaster in a small town may remember what was on a change of address form even if it's no longer on file.

In a pinch, try contacting the mail carrier on the route that included the old address. With urban mail carriers this may not work, because they often don't stay on the same route very long and don't know the people well. Mail carriers on rural and small town routes, however, are often quite friendly, know the people on their routes personally, and are regarded by investigators as excellent sources if properly approached.

Although change of address notices are only kept at the local post office for a limited period, the data is not destroyed. It goes into the U.S. Postal Service's National Change of Address System, a magnetic tape database that contains permanent change of address data on more than 25 million relocating customers. Although you cannot just call up Washington and get this information, the Postal Service does sell the tapes and frequent updates to more than a dozen mailing list companies, who use it to keep their lists up-to-date. Information brokers then buy the mailing list databases, add their own software, and offer skip tracers and other customers access to the National Change of Address System!

## 4.7 City and Crisscross Directories

If the above methods fail, your next step will be the city directory or crisscross directory. As well as the current edition, you will probably need previous ones, which may be on microform at your library or possibly can be accessed through the directory publisher's subscriber call-up number. Essentially there are three methods: first, find out who lived with subject at his last listed household, track them down at their current addresses, and persuade them to give you subject's current address (or at least, a more recent address); second, contact the current residents of subject's former house and see if they know anything; third, question subject's former neighbors. Note that these techniques are valuable not only in finding people but also in compiling detailed background or security clearance reports and in writing biographies of celebrities.

### Finding Former Household Members
If the given locality was covered by a city directory at the time subject lived there, the directory may list all members of subject's former household and provide information that will help find them today, e.g., their

occupations or former places of employment. If the given locality was not covered by a city directory, you still can get the names of some of the former household members via the crisscross directory:

- The numerical section in back editions of the local crisscross directory can be used to find the multiple listings (both same-surname and different-surname) under subject's former phone number; in addition, it can be used to spot instances of duplicate service (two people sharing the same phone number but with lines in different apartments or houses within the same telephone exchange) and instances in which a person who shared the phone chose to list an address other than the actual household's (for instance, a P.O. box) or no address at all.

- The street listings in the crisscross directory back issues can be used to find the name of anyone in the household who at any point had his own separate telephone account with a listed number (unless that person chose not to give an address or gave an alternate address in the listings).

Separate phone numbers or multiple directory listings for a single phone number may occur under the following circumstances (among others): households with teenagers who have active social lives, married women retaining their premarriage names for professional reasons, roommates or friends sharing a household, adults living with parents (or vice versa), renters or sublessees who live in the home while the owner or primary tenant is away for an extended period, and businesses being operated out of the home (note that 20 million Americans operate part-time or full-time home businesses today). In addition, a separate phone account under a different surname at a private home may be that of the family that rents the basement apartment. Or the different-surname listing or different-surname separate account may be that of a child from either spouse's previous marriage who keeps the name of the previous spouse. (A person with the latter surname listed at another address might indeed be this previous spouse and might know subject's current whereabouts.)

As crisscross directories do not include people with an unlisted number (or people with no telephone at all)—or household members with no separate listing or separate account—you may want to supplement your search with other types of street listings that do not have this limitation. Your board of elections may maintain street listings of all registered voters (in New York City, there is a computer printout volume for each Assembly District). These may list any member of a household who is both old enough to vote and has bothered to register. Likewise, if you can access the state Department of Motor Vehicles database on your computer (or can get an information broker to do so), you might get a printout of every licensed driver or car owner at the given address. This might also be done with parking violations bureau indexes.

## The New Residents of Subject's Former Home

If no one was listed at the previous address except subject—or if none of the former household members will talk—it's time to go to the second stage of your crisscross/city directory search. Obtain from the street directory the name and phone number of the new occupant of subject's previous home and also of the closest neighbors on the block. (If you are using the public library's crisscross or city directory, photocopy all relevant pages before beginning your calls.)

First check out the new occupants of subject's former home, since (1) they may be renting or subleasing from subject (with a new telephone number in their own name of course); (2) they may have bought the house or co-op apartment from subject and have a current address for him or her; (3) they may have bought the house from a third party who acquired it from subject and knows subject's current address; (4) they may be renting from the same landlord from whom your subject rented. In the latter case, ask them to give you the landlord's phone number (or the number of the rental office or managing agent, in the case of an apartment house). If subject skipped out owing a lot of rent, the landlord may provide you with information from subject's rental application form—or from a credit check the landlord once ran on subject—in hopes you will reciprocate by giving landlord the new address when you get it.

## New Residents of Subject's Former High-Rise Apartment

Often the new tenant of an apartment will know nothing about his or her predecessor. Sometimes, however, the new tenant is a friend of the previous one and obtained the apartment on the latter's recommendation. Indeed the current tenant may be subleasing from the previous one. Furthermore, if the apartment is a co-op or condo, the new tenant may be directly renting from the old tenant, who is now the apartment owner (of course in this case you can probably get his or her new address from the co-op board or the building management).

Crisscross directories often do not give the apartment numbers but only the phone numbers for residents of large apartment buildings. The quickest way to find out which tenant now occupies subject's former apartment is to ask the super (the latter usually lives in the building—just call any tenant at random and ask for the super's name). If the super won't help you, go back to the crisscross directory, which tells the number of years each telephone number has been listed. Call up all tenants who moved in around the time subject moved out. (Note that this will only work if the new tenant of subject's former apartment has a listed phone number.)

## Former Neighbors

Next try subject's former neighbors. The latest crisscross directory will tell how many years each telephone number has been listed at a given address, so you can figure out which neighbors were around at the same time as subject. In approaching former neighbors, start with those on each side of subject's former house and also those directly across the street (the latter might be the most likely to have noticed the name of the moving van company). If these all moved in after subject moved out, start with the nearest neighbors who resided there at the same time as subject. Note that in some neighborhoods people will as likely be friends with back-fence neighbors as with the neighbors next door. Since the back-fence neighbors will be listed on another street in the crisscross directory, you will usually need a city map to figure out who to call. If you encounter a gossipy former neighbor, let them talk awhile. You never know what clues you'll pick up. For instance, the neighbor may recall that subject had a serious weight problem. Later, when you get a tip that subject is living in Portland, you might want to call weight loss clinics in the Portland area.

If most of the current residents on subject's former block are people who moved in after he or she left (a frequent problem in neighborhoods that are rapidly changing their ethnic character), you will have to use a back-issue crisscross directory to identify residents who lived on the block at the same time as subject. These people can then be looked up in the phone book at their current addresses.

### Neighborhood Stores

Look in the street directory to see which stores in the area were in business at the time subject lived there. Based on what you've learned about subject, decide which are most likely to remember him or her. If subject had pets, you might check with the nearest vet's office. If subject's wife is especially fashion-conscious, check neighborhood boutiques. Other possibilities might include local garages, video stores, bookstores. Business establishments that are centers of local gossip, such as beauty parlors, should always be called.

If subject left town owing money to a local merchant, there are two possibilities. First, subject later paid by mail, in which case the merchant may remember the address on the check or the return address on the envelope. Second, subject never paid up, in which case the merchant may cooperate by making inquiries for you among customers and other store owners.

Checking with merchants may be a long shot in areas where most shopping is done in vast malls twenty miles from home; still, there is usually an equivalent of the old corner store (or country store) for neighborhood convenience shopping.

## 4.8 If the Number is Unlisted

Private investigators often have informants inside the phone company who will provide them with unlisted numbers. News reporters sometimes get a police officer or local politician to obtain a number for them. However, there are many other ways to get an unlisted number, some of them almost ridiculously simple:

- When you call Directory Assistance, always ask for any business listing in a person's name as well as the residential listing. Often the former will be listed when the latter is not, perhaps even at the same residential address. (If you call the business number during off-hours, a recording may give you the unlisted residential number.)

- Search the non-credit "header" information in credit-reporting agency databases, which in some cases will include a subject's unlisted phone number. (See 2.6.)

- Dial "0" and ask for the supervisor. Tell him or her it's an emergency and that you must contact the person with the unlisted number. The phone company will call the person and pass on your name and number. Of course if it's *not* an emergency the person will be extremely annoyed, and thus probably refuse to cooperate with your investigation.

- If Directory Assistance tells you the number is unlisted, always look in the phone book. Subject may have obtained his or her unlisted number very recently. If subject has remained at the same address, you may find that the previously listed number is, in fact, the very number that is now unlisted (it costs less to unlist the old listed number than to get a new unlisted number). If not—and if subject has moved to a new address—you may get a recording on the old number providing the new one. (If the recording merely says the old number has been disconnected, try again in a few days—the telephone company may simply not yet have connected the new phone.)

- If there is no listing for subject in the current phone directory, look in your collection of back issues until you find a listing. The number unlisted for the past ten years may have originally been listed.

- Call listed numbers of all persons with the same surname in the white pages. If subject has several relatives nearby (especially in a smaller population center), this often produces quick results. You just assume you have the right number and say, "I'd like to speak to Marvin Klenetsky." They say, "Oh, you have the wrong Klenetsky; he's at . . ." and they give you the number automatically. Unlisted numbers are not exactly a deep secret; often the relative doesn't even know the number he or she is giving out is unlisted.

- The unlisted number may be found in a city directory, either alphabetically or in a street-listing format. City directories, available at your local chamber of commerce or public library, compile their information on each household by the survey method (unlike crisscross directories, which compile their information from phone company directories). The survey taker will ask the member of the household contacted to provide, among other things, the household's phone number. Only a very small percentage refuse to give out the number, even if it is unlisted in the white pages. (If they do refuse, the city directory will at least tell you the address of the family and possibly the place of employment of the head of household.) In most large cities, city directories have been discontinued for many years. However, if you know your subject has lived in the same home for decades but has never been in the phone book, you might want to check the final edition of the city directory on microform at your public library. Subject's current unlisted number—its unlisted status maintained all these years—may be in that final 1962 edition. Of course a person who has lived in one spot so long will probably be very well-known to neighbors; thus, you might find it easier to call longtime neighbors listed in the current street directory to obtain the number.

- Check public records at the county courthouse or town hall (in some localities and for some records this can be done over the phone). The unlisted number may be included on, say, a license application.

- Subject may have a weekend or vacation home at the seashore or in nearby mountains. Although keeping unlisted the phone number of his or her main (weekday) residence, subject may not have thought to do likewise with the weekend phone.

- Some corporations distribute internal phone directories to their employees giving the home addresses and phone numbers of each. These directories will often include phone numbers and addresses not listed in the telephone company white pages. Thus, if you know where your subject works or used to work, you might try to get the number and address from another past or present company employee who has a copy of the directory. Note that such directories are frequently obtained by personnel agency headhunters, who use them to call around and see if anybody's interested in moving to a new job.

# 4.9 If You Have a Number and Name, but No Address

Your subject may have chosen to list his or her phone number in the white pages without an address or with only a partial address (street but no number) or with a box number. Often the full street address can be obtained simply by looking in a back-issue phone directory. Generally, if the number listed in the old directory corresponds to the number in the new, the address in the old will be the same as that deleted from the new (unless the person moved within the same telephone exchange and asked to keep their previous number). You can easily confirm if subject is still at the old address by calling Directory Assistance. Give the operator the surname of your subject and the "old" address—specify that you *only* want the party at that address. If the number you obtain is the one currently listed without an address, ask the operator to confirm that it coincides with the address you provided.

If subject has moved to a new address, you can determine the general location via the first three digits of the telephone number, which designate a telephone exchange zone covering a particular town or towns (or section of a city) within the given area code. If a list of the exchange zones and corresponding numbers is not included in your telephone directory, you can obtain it from the phone company. This information may help to narrow your search: If you want to contact Susan Smith, and you believe she lives in a certain town in Nassau County on Long Island, then if you find an "S. Smith" listed without an address, but with a zone number for eastern Suffolk County, you can assign this listing a low priority.

In some localities, Directory Assistance will tell you which town the first three digits of a phone number designate.

## 4.10 If You Have a Number, but No Name or Address

For news reporters this is important when they start getting calls from mysterious people trying to pump them for information. If the person gives you a phone number (or if you get his or her number via Caller ID), look in the numerical (reverse) listing section of the appropriate crisscross or city directory. In some cities, you can call the public library telephone reference service and they'll look it up for you. If the public library doesn't have the directory, try the local chamber of commerce.

If the number is in Chicago, call the Chicago phone company's reverse directory at (312) 796-9600. They'll give you both the name and address of the customer.

## 4.11 Tracking a Person Via His or Her Family and Former Friends

Most people have a hometown (or, at least, a town or city where they spent considerable periods of time during their early years), and most stay in contact throughout adult life with one or more members of their family of origin, who often have remained in the hometown. Likewise, most people establish families of their own; these family units may not remain intact, but even the worst reprobate usually maintains some kind of contact with his or her children.

## Finding a Person's Hometown and Members of His or Her Family of Origin

When attempting to track subject through former next-door neighbors (see above), always ask if they recall where subject grew up. If they don't, you can find this information or clues to it in various records. For instance, high-school graduation information—and sometimes, place of birth—are included on job applications. Credit-reporting agency databases will often have the place of birth and may have a string of addresses for subject going back to his or her hometown days. In addition, you can try old Selective Service records, military discharge papers, and driving (moving violations) records. Subject's Social Security Number may be your best clue—it is coded to reveal the state of issuance (see 7.4), which probably is a state in which subject lived as a teenager or young adult.

If subject has listed himself as "John R. Doe, Jr." or John R. Doe III," you could search national or regional telephone databases (see 4.4) for the father or grandfather, getting printouts of all listings under the names "John R. Doe," "John R. Doe, Jr.," "J.R. Doe," "J.R. Doe, Jr.," and perhaps other permutations. In addition, if subject has a very unusual surname, or a very unusual spelling of a common surname, you might want to get printouts of every listing for this surname from the regional telephone company databases or from Phone*File. If you know subject's home state, you might want a printout of everyone in that state with the same surname even if it's a rather common one.

When you call the same-surname people cold, begin with those in subject's hometown if you know it. Start with persons with the same first name and middle initial (subject might be a "Jr." without your knowing it). This method is not guaranteed, but there is a good chance that, with a few calls, you will find someone who is at least a cousin of subject and will be able to steer you to the immediate family.

If you know subject's mother's maiden name (sometimes found on loan applications or in credit-reporting agency records), you might also call local people with that surname to find maternal grandparents, aunts, uncles, and cousins.

You may find that subject's family left the area long ago and that they are apparently scattered throughout the country. Nevertheless, the hometown records may give you their names and perhaps clues to where they live now. Back-issue city directories may tell you the names of everyone who lived in the household while subject was growing up—his or her siblings, parents, perhaps a grandparent. (The city directory may also have information from that period on the households of other relatives.) As described above, back-issue crisscross directories and white pages can also be used to gather the names of at least some former household members.

### Society-Page, Obituary, and Other Clippings

You might also obtain clips filed under subject's name at the local newspaper morgue (see 6.9), as well as clips for the other household members listed in the city directory and even for all local people with subject's surname. The clips may include information about the arrest of one family member and the winning of a lottery by another, but you will mainly be looking for an overview of the family. A society page article about subject's marriage will provide the names of his or her parents (possibly including subject's mother's maiden name) and other family members. An article about a sister's wedding will give her married name, and the list of guests may fill in gaps in your own list of family members. An article about the parents' silver wedding anniversary—or a notice of their divorce—may also be useful. In addition, an obituary notice for a parent or other family member may provide the names of all siblings (including the married names of all married sisters) and their cities of residence at the time of the funeral. It may also list subject's surviving aunts and uncles on the deceased parent's side.

If a daily newspaper does not have a general index, it may at least keep a card file of the obituaries it has printed. Large metropolitan dailies generally do not print obituaries on ordinary citizens, but you may find a brief mention in the paper's vital statistics column; also, an ordinary person's death may be mentioned as a news item if the death was a murder, auto accident, drowning, etc. Obtaining the date of death from such news items (or from the Social Security Death Index—see below), you could then look for an obituary in a suburban daily or an ethnic or neighborhood weekly. The city directory will have told you the occupation of the father and/or mother and possibly where they worked; you could thus look for an obituary in a local or statewide trade, professional, or labor union newspaper or in the house organ of the company where the deceased worked. Finally you could look for a death notice placed in any of the above publications (including the metro daily) by grieving family members, friends, or co-workers and listing the names and cities of residence of the survivors; or a thank-you notice from the family in a newspaper issue following the funeral.

### Microfiche Records and Indexes at Local Genealogy Centers

You will most easily find what you need in old newspaper files if you first get the dates of family members' births, marriages, and deaths from county or state vital statistics indexes. Often microfiche of these indexes is available at the county historical society or the public library. In many localities, the best resource is the local Family History Center (FHC) run by the Mormons. These centers, located throughout the country, are listed in the white pages under Church of Jesus Christ of Latter-day Saints. Each FHC has the microfiche Family History Library Catalog, which covers the holdings of the Mormons' Family History Library in Salt Lake City. The catalog gives a breakdown by locality of 1.7 million microfilm rolls of local records, including birth, marriage, and death indexes; wills, deeds, and land records; and church records (e.g., baptismal records) from churches of all denominations. The local FHC will have the microfilm records for its own locality; microfilm for other localities can be borrowed from Salt Lake City for your viewing—it's one way to search old courthouse records in Oregon if you're stuck in New Jersey. The catalog also lists alphabetically by surname a comprehensive list of family name histories and genealogies (this list can be supplemented by *Genealogies in the Library of Congress* and *Complement to Genealogies in the Library of Congress*).

You will also find at the local FHC the International Genealogical Index (IGI), which lists more than 147 million living and deceased people in the United States and overseas, and the Social Security Death Index (both are on CD-ROM and are thus cross-searchable). The latter includes the death records of 39.5 million U.S. Social Security registrants, most of whom died between 1962 and the present. It is alphabetical by surname and tells the date of death, the date of birth, the place the SSN was issued, and the place of residence at time of death. (Once you get the SSN of the deceased, you can get more information from the Social Security Administration.)

### Biographical Dictionaries and Obituary Indexes

Another quick way to get information about subject's family is to look in the same-surname listings in the *Biography and Genealogy Master Index* (*BGMI*) (see 5.2). Your subject may be an obscure person, but he or she may have a notable father, mother, or sibling who is listed in a half dozen biographical dictionaries. All it takes is one notable in the family, and you've got a wealth of information—including the hometown location—for family backgrounding purposes. (If a parent is listed, then the entry may include the names and dates of birth of all children, as well as the name of the mother and/or stepmother.) For obituaries of notable persons, see the *BGMI* and the *New York Times Obituary Index* (the latter includes 353,000 names from 1858 to 1968). For full indexing of *Times* obituaries up to the present—and for indexing of *Times* articles on deaths from accident or murder—see the *Personal Name Index to the New York Times Index* (described in 6.8).

### Probate Court Files

Another way of getting information about subject's family is through the local probate court, which administers estates, trusts, and guardianships and includes the office of the registrar of wills. Here you may find detailed information about the family and its finances. Also check if the children ever filed a motion for conservatorship over an elderly parent's affairs (this is handled by state district court rather than probate court in some jurisdictions). In an investigation several years ago, I discovered such a court action in New York State Supreme Court, listed under my subject's name as plaintiff. In the microfilm records, I found the names and addresses of all six of subject's siblings (his co-petitioners), most of them in the metropolitan area. This proved invaluable in tracing subject's real estate dealings under straw names.

## Tracing Subject Through His or Her Ex-Spouses

The maiden names of subject's current and former spouses and the names and addresses of current and former in-laws may be found in the marriage column of a local paper. In addition, you may discover the names of spouses and ex-spouses in back-edition city, crisscross, and white-pages directories, as well as in marriage and divorce indexes (your county court-house may have a divorce and annulment index, for instance, listing cases by plaintiff and year) and in the "header" material on subject's file in a credit reporting agency database.

If a male subject's first wife has not remarried, she may be listed in the phone directory under either her maiden name or her ex-husband's surname. If she has remarried and is using her new husband's name, but you don't know what it is, you can often contact her through her former divorce attorney, whose name and address will be in the court file. You may also try to contact her through her parents, whose names were in the hometown newspaper marriage clippings. If she and her new husband have a teenage child from her previous marriage living with them, that child may have a separate telephone listing under subject's surname.

If you are having trouble finding an ex-wife's maiden name, check the following possibilities:

1. She and your subject may have lived together before they were married. During that period she may have had a separate listing for their telephone under her maiden name or a previous married name. Or subject may have moved in with her after their marriage in an apartment where the phone was previously in her maiden name.

2. A teenage child may have adopted a hyphenated name (mother's maiden name/father's surname) even if her mother never did so. Look for a separate listing on the previous family phone or a separate phone at that address in the child's name.

3. Couples may use hyphenated names temporarily or on isolated occasions for various reasons; you may spot some public record in which subject and/or his wife did so.

In searching for a male subject's ex-wife don't be blinded by male chauvinist assumptions. Although subject is an obscure person, his ex-wife (indeed, his current wife) may have become quite a successful person. Look for her under both her married and maiden names in *BGMI* and in *Biography Index* (if this search succeeds, you will get from the cited biographical dictionary entry the names of her children). If she is a professional person, you may find her in one or more professional or trade directories or rosters (see 5.5).

## Tracing Subject Through His or Her Children

If subject's children are grown, check for current local telephone listings under their names. Note that after the splitup of their parents and the remarriage of their mother, these children might have adopted their stepfather's name, their mother's maiden name, a hyphenated version of their mother's/stepfather's name, or a hyphenated version of their mother's/real father's name (see 7.1). If you don't find them in the phone book, try other search techniques described above.

## Contacting Former Spouses, Relatives, etc.

Note that a subject hiding from the police or from bill collectors often has left a string of legal or common-law marriages and children in his or her psychopathic wake. However tragic this might be, it makes the investigator's job easier: Bitter ex-spouses are excellent sources of information, especially if subject has skipped out on child-support payments or just skipped out period, leaving his or her erstwhile partner to raise a houseful of kids alone. However, not all marital breakups are like this, and some divorced couples even remain close pals. The ex-wife or ex-husband you contact may be quite loyal to subject. Even if not, the "ex" may refrain for the sake of the children from providing any information harmful to his or her former spouse.

In contacting ANY relative, ex-spouse, son, or daughter of subject, be careful what you say. If your reason for finding subject is benign, and if subject is not in hiding, a relative may either give you subject's number or pass along a message that you wish to speak with him or her. But if subject is hiding out, his or her loved ones may refuse to talk unless you employ a ruse.

Skip tracers and bounty hunters will frequently call up the phone company using the name of, say, subject's brother, whom they suspect is in close contact with subject. The putative brother complains about alleged mistakes on his latest long-distance bill and asks the phone company representative to read out each number called and the date of each call from the computer record so he can verify them against his own printed bill. He copies down each number and later reverses them via an online crisscross database. This gives him the addresses and names of the telephone subscribers whom the brother called. One of these may be subject living under a false identity, or a friend with whom subject is staying. If the bill includes a long-distance collect call from a pay phone in a city 2,000 miles from the brother's house, it may suggest to the skip tracer that subject is living in that city, and possibly lives, works, or hangs out in a bar within a few blocks of the phone booth.

The skip tracer then calls any number or numbers he suspects might be subject's. If there is an answering machine, he tapes the voice and then plays it back to someone who knows subject's voice. If it is not subject's, the skip tracer makes additional calls to the same number at various times of the day and evening, pretending to be a telemarketer while taping the voices of the various people who answer the phone in person. He then plays back these additional recordings to his source. (This has to be done carefully if subject is a fugitive from the law; any phone calls that seem vaguely suspicious are likely to set him or her running again.)

### Subject's Close Friends of Years Past

If you can't find family members, you can still possibly trace subject through childhood or early adulthood friends with whom he or she might have kept in touch. High school or college yearbooks will give the names of classmates; by noting who was on the same team or in the same fraternity or sorority with subject, you can compile a list of classmates most likely to have kept in touch with him or her. (The college classmates will be easily traceable through the alumni directory; for high school classmates, see the telephone directory of the town in question.) A clipping from the local newspaper's society page will give you the names of the maid of honor and other bridesmaids—and the best man—at subject's wedding. In addition, the clipping may tell where both bride and groom were employed at the time of the wedding; employees at those firms may still be in touch with them.

## 4.12 Local Churches and Other Organizations

Check with the church secretaries at local churches of subject's denomination—they may have records regarding a transfer of membership. If you want to contact members of subject's former congregation who might have known him or her well, ask the church office to send you a copy of the membership mailing list. Churches will often tell you just about anything about congregation members except their sins.

Make similar inquiries with subject's former civic club or fraternal lodge and any other organizations to which subject or other members of the household belonged. You might even check with the scoutmaster of subject's son's former troop.

## 4.13 Former Employers and Co-Workers

Subject's former place of employment can sometimes be found in a listing for subject in a back-edition city directory or in the Federal Election Commission's online Contributor Search System (see 10.10). Try calling the firm's personnel office. If they can't help you, go back to the city directory and see if anyone on subject's former block worked at the same firm during the same time period. Although such a search might be tedious, it could turn up someone who knew subject fairly well. Even if the former co-worker doesn't have subject's current address he or she may at least remember some potentially useful personal trivia, e.g., whether or not subject has a dog, and the name and breed of the dog. (This can be useful in identifying subject if he or she is living under an assumed name somewhere with an altered physical appearance.)

Sometimes the city directory or the FEC records only tell you subject's occupation without giving any place of employment. In such cases, you can look in the local yellow pages and *Thomas Register of American Manufacturers* to find the names of area firms most likely to employ people in subject's job category.

## 4.14 Caller ID

If you are calling around to find people who don't want to be found, they may hear about you from a former neighbor and become curious or angry. They may call your number just to hear what your voice sounds like, and then hang up without saying anything. They may call pretending to be someone else and try to pump you for information. Or they may call anonymously with threats or obscenities or heavy breathing. Caller ID will tell you the number from which the call was made. If you call back immediately and ask for subject, he or she may break down and talk to you. If you reverse the number in the crisscross directory to get the address and name, you may find the number is subject's although listed in a fictitious name. (For more on Caller ID, see 3.8.)

## 4.15 Professional Directories, Alumni Directories, Etc.

Professional and trade organizations, college alumni associations, and the reference book industry publish thousands of directories each year for the purpose of helping people with mutual interests find or keep in touch with one another. These works give current or recent addresses for tens of millions of Americans.

To use this resource you will need a little background information on your subject; for instance, his or her occupation or profession, college alma mater, etc.

Look in *Directories in Print* and *City & State Directories in Print*, checking every heading that might pertain to your subject. Then call your public library's telephone reference service and ask them to look in the most likely volumes. If your library doesn't have a certain volume, call a library in another city.

Is your subject a full-time or part-time college teacher? The three-volume *National Faculty Directory* lists almost 600,000 college teachers at nearly 3,700 American and Canadian junior colleges, colleges, and universities. Always check the annual supplement, which includes tens of thousands of new and updated listings. If subject has retired or quit teaching, you may still find him or her in a back issue.

Is subject a lawyer? The sixteen-volume *Martindale-Hubbell Law Directory* gives the business addresses of over 800,000 lawyers and law firms in the United States and Canada, divided according to town, city, state, and province.

Is subject a physician? The American Medical Association (AMA) publishes the *American Medical Directory*, a biennial register of all AMA members in the U.S. and Canada.

The variety of such directories is extraordinary, covering even very obscure professions and trades; I suggest that you spend some time browsing through *Directories in Print* and its sister volume to get an idea of the possibilities.

If the comprehensive national directory for a given profession does not include your subject's name, look in *City & State Directories in Print* for the names of local or regional directories and also check specialized-focus directories for subject's profession. Many attorneys are not listed in the massive *Martindale-Hubbell* (especially those who are no longer in practice), but if subject is a woman you may find her in the *Directory of Women Law Graduates and Attorneys in the U.S.A.* (more than 25,000 names).

America's college alumni associations are probably the largest source of directory-type listings. Virtually every college and university has an alumni directory that gives the names and current addresses of most living alumni as well as annual listings of recently deceased alumni. Alumni associations need these rosters for fundraising purposes, and often pay search agencies to find missing alumni and keep the rosters up to date.

If you know which college subject attended, simply call the alumni office on campus and ask them to look up subject's address in the latest directory. If they don't have a recent address, you might try the national office of subject's fraternity or sorority—its membership directory might have the current address. (To find out the name of subject's fraternity or sorority, you'll have to persuade the librarian at the campus archives to dig out the yearbook for subject's senior year; see 11.1.)

Note that many alumni directories list not only graduates but matriculants who dropped out, flunked out, or transferred. Thus, if Jane Doe is not listed (or is listed at an out-of-date address) in the alumni directory of the state university from which she received her B.A., you may still find her current address (or a less out-of-date former address) in the directory of the women's college she attended during her first two undergraduate years. And if she's not in either of these, she may be in the graduate student alumni directory of the urban university where she studied for her M.A. Alumni associations vary widely in their commitment to keeping their mailing lists up-to-date; thus, the more colleges or universities subject has attended, the greater your chances of finding a current address.

If you find that subject's current address is not in the latest published directory, the alumni office might be willing to check its update files or its mailing list for you.

## 4.16 Electronic Yellow Pages

Dun's Electronic Business Directory, available on DIALOG, provides directory information on over 8.2 million businesses and professionals throughout the United States. Compiled from almost 5,000 telephone directories and updated semi-annually, its "Professionals Directory" includes almost 2 million listings.

## 4.17 Membership Rosters

If the above methods fail, cast a wider net to the vast number of trade, professional, labor, charitable, religious, hobby, or sports organizations that maintain rosters of their members (although perhaps not in the form of a directory sold to the public) and mailing lists of financial supporters.

Two library reference works that will orient you to the vast world of rosters and mailing lists are the *Encyclopedia of Associations* and the *National Trade and Professional Associations of the U.S.* These two books are indexed in a variety of ways. You can easily find the names of associations to which your subject might belong.

Professional groups such as the state or county bar associations may provide the current business addresses of members to the general public. Note that local associations or branches are often more helpful with inquiries about local members than are the national or statewide organizations.

If you want to find any ex-member of Congress, call the U.S. Association of Former Members of Congress at (202) 332-3532. They will tell you his or her present address and phone number.

Even if an organization does not give out members' addresses, they may be willing to forward a letter in which you ask subject to contact you (see 4.29).

## 4.18 Subscription Lists

A former neighbor of subject recalls that subject played the fiddle and often attended bluegrass conventions. You go to the *Standard Periodicals Directory* and look under the subject category, "Music and Music Trades," where you find the publication *American Fiddler News*. If you can get someone in the subscription department to check the mailing list, it might include subject's current address. (Also, you might ask them to check the membership roster of the American Old Times Fiddlers Association, which publishes *American Fiddler News*.)

## 4.19 Licensing Agencies and Certification Boards

In any locality of the United States, the city, county, and state governments license hundreds of professions, trades, and types of business activity. The roster of license holders is a good place to search for your subject, if you can figure out which licenses he or she is likely to have. For a full discussion of licenses, see 10.11 and 10.12.

## 4.20 Department of Motor Vehicles

An abstract of a person's driver's license (see 10.2) will give you his or her address at the time of license issuance or renewal. To access this license information by mail or over the phone, you may need to provide subject's social security number. (Note that in some states the Department of Motor Vehicles offers this information online, with cross-search capabilities, to any computer user who cares to subscribe; you can also go through an information broker.)

If you think your subject has moved out of state, the Department of Motor Vehicles may have records of expired and transferred licenses telling to which state subject's records were sent when he or she applied for a new license. You can then get from the latter state an abstract of the new license which will include a more recent address (although, if subject is playing hard-to-find, the new address may simply be a mail drop).

Note that many people who don't know how to drive or whose licenses have been revoked will apply to the Department of Motor Vehicles for a non-driver's State I.D. card. The information on their application, including address and SSN, may be publicly available in your state.

## 4.21 Abandoned Property Lists

In New York State, about six million names and addresses are on these lists, maintained by the state's Bureau of Unclaimed Property. Updates are published twice a year and are available at municipal offices and public libraries. Although the majority of addresses are outdated, the address listed for your subject may be newer than the one with which you began your search.

## 4.22 Miscellaneous Local Government Records

If you've narrowed your search to a single locality, you can check various public records at the county clerk's office, the municipal building, the register of deeds office, etc. Your options may include the tax assessment rolls, the county water commission's roster of accounts, and the "grand lists" (the latter are lists of persons paying taxes on valuable personal property such as autos, boats, and airplanes). You also might check the judgment docket, the wage garnishment index, the Uniform Commercial Code (UCC) filings, the federal tax lien index, the courthouse plaintiff/defendant indexes, and the roster of municipal employees.

If your subject is a low-income person, the voter registration lists may be your best bet. (The Democratic Party registers millions of low-income workers and welfare recipients every Presidential election year.) Another way to find low- and moderate-income persons is by examining the docket index for the county or municipal civil court (claims for under $10,000). Certain plaintiffs (such as municipal or voluntary hospitals, finance companies, etc.) will each have cases against hundreds, if not thousands, of ordinary people. If the index only lists cases by plaintiff, the defendants may be listed in alphabetical order under the plaintiff's name. Also, if a large percentage of your city's residents are apartment renters, try the housing court index. In New York City, landlords file eviction proceedings against over 400,000 tenants each year. Look under subject's name and also his or her spouse's name (including maiden name) or live-in lover's name in the computerized index. To search housing court indexes for an entire metro area or state, check with the tenant screening databases used by landlords. (A certain company in New Jersey charges $10 per search for eviction data drawn from twenty-one local courts, while a California company claims to have records on more than two million tenants.)

The scofflaw index is a marvelous tool for finding people of all socio-economic brackets. In New York City, every car owner who has an unpaid city parking ticket is listed, together with the address on his or her car registration at the time the ticket was issued. The Parking Violations Bureau has computer terminals at which these indexes can be easily searched. In addition, I have often used, at the county clerk's office in Manhattan, the multivolume computer-printout indexes, which include separate volumes for scofflaws who received their tickets in New York City but reside in New Jersey, Connecticut, etc. There are back sets available, so that even if subject has paid up you can find a listing.

The backfile of scofflaw index volumes, if such exist in your city, may be located in the municipal archives. Although the addresses, even in the current index, may be years out of date, they at least will give you a place to start (and they may include addresses you would never have found in the telephone or criss-cross directory).

For journalists in a central city, scofflaw indexes are a quick way to locate residents of the suburbs and exurbs. A large percentage of these people work in the city, and many others come into the city to shop or dine, getting parking tickets in the process.

Skip tracers recommend looking at the county roster of dog owners. In many localities you can obtain a copy of the entire roster for a small fee.

A good rule of thumb is: Begin your search with the records systems that list the largest numbers of local residents alphabetically on a citywide or countywide basis with addresses updated annually (the tax rolls, for instance). Work your way down to the narrower and often harder-to-search records systems.

For further description of the various courthouse and municipal records, and their larger uses in compiling background reports, see Chapters Eight, Nine, and Ten.

## 4.23 Trailer Rental and Moving Companies

If your subject has recently moved, ask the neighbors (or the doorman or super, if it's an apartment house) what they remember. Did subject hire a small van or U-Haul-type trailer and do the moving with the help of friends? Or did he or she hire professional movers? Did the movers arrive in a panel truck or in a large moving van? If no one remembers the rental company name on the trailer or the movers' name on the side of the truck, you'll have to start calling the various rental companies and movers listed in the local yellow pages.

## 4.24 Mail Drops

If you think you've finally found subject but the supposed home address turns out to be a suite in an office building, subject is probably using a mail drop. He or she may not even be in the same city: The mail drop could be forwarding subject's mail to just about anywhere in the world. Your first step is to check the *Directory of U.S. Mail Drops* (see bibliography), which includes over 1,200 such companies. If subject's alleged address is not included there, look in the yellow pages to see if any private mail service is listed at that address. Be aware that a small number of mail drops are actually run by, or sell their forwarding address lists to, firms specializing in skip tracing. (The latter do "reverse traces" on the mail drop customers, find out who they owe money to, and then sell the forwarding addresses to the creditors.)

# 4.25 Locating Present, Former, or Retired Military Personnel

Each military branch has a locator service that will tell you the military unit and installation to which subject is currently assigned. Provide subject's full name, date of birth, SSN, and any other basic data you have.

To find those on active duty in the Army, write: U.S. Army World Wide Locator, EREC, Ft. Benjamin Harrison, IN 46249. For those in the Army Reserve or Inactive Reserve—and those retired from U.S. Army active duty, the Army Reserve, or the Army National Guard—write: U.S. Army Reserve Personnel Center, 9700 Page Boulevard, St. Louis, MO 63132. For members of the Army National Guard, write the State Adjutant General of the given state.

For those on active duty in the Navy or in the Navy Active Reserve, write: Chief of Naval Personnel, PERS-312D, Washington, DC 20370 (the Navy will provide the U.S. land-based unit location only). For personnel retired from active duty or from the Navy Reserve, and those in the Inactive Reserve, write: U.S. Navy Reserve Personnel Center, Code 12, 4400 Dauphine Street, New Orleans, LA 70149.

For those on active duty in the Air Force or in the Air Force Reserve or Air National Guard, or retired, write: U.S. Air Force World Wide Locator, 9504-IH, 37 North Street, San Antonio, TX 78233. (The Air Force will not provide the location of overseas personnel.)

For those on Active Duty in the Marines or in the Marines Selected Reserve, write: Commandant, U.S. Marine Corps, MMRB-10, Locator Service, Building 2008, Quantico, VA 22134. For Individual Ready Reserve and Fleet MC Reserve/Inactive Reserve, write: U.S. Marine Corps Reserve Support Center, 10950 El Monte, Overland Park, KS 66211. For retired Marines, write: Commandant, U.S. Marine Corps, MMSR-6, Washington, DC 20380.

For those on active duty in the Coast Guard or in the Coast Guard Reserve, write: Commandant, U.S. Coast Guard, G-PE, Locator Service, 2100 2nd Street S.W., Washington, DC 20593. For those retired from either Active Duty or the Reserve, write: Retired Military Affairs Branch, G-PS-1, U.S. Coast Guard, Washington, DC 20593.

Once you know the base or post at which subject is stationed, you can check with the locator at that facility, who may provide subject's unit or ship assignment and work phone number. A directory of military bases and other military facilities in the United States, with zip codes, telephone information numbers, and base/post locator numbers, is included in Lt. Col. Richard S. Johnson's *How to Locate Anyone Who Is or Has Been in the Military: Armed Forces Locator Directory* (see Bibliography).

For overseas personnel, the servicewide locators provide unit location by post office zip code only. Johnson's book provides a list of zip codes matched to overseas base/post. The locations can also be identified via the U.S. Postal Service Zip Code Directory.

The service locators will forward letters to current armed forces members at home or abroad; however, the addressee is not obliged to answer you. To take advantage of this service you must provide the basic identifying information described above. If you already know the base/post where subject is assigned, you can send the letter c/o the local base/post locator.

Records on all personnel discharged from Active Duty or Reserves (all services) are kept at the National Personnel Records Center (NPRC), 9700 Page Boulevard, St. Louis, MO 63132. The NPRC will not give out addresses. Under a few circumstances (such as a financial institution attempting to collect a debt) the NPRC will forward correspondence to the last known address. The NPRC will not forward correspondence for persons seeking lost family members; for this you must go to the government locator services described in 4.28.

You can locate some war veterans through a private organization, Information Up, which has 50,000 addresses in its databases and access to many more. Write: Information Up, Veterans Locator and Resource Center, 4614 Hamlet Place, Madison, WI 53714.

For other methods of locating discharged military personnel, see Johnson's book, which should be in every investigator's library.

## 4.26 Locating Licensed Pilots

You can obtain the address of a Federal Aviation Administration (FAA) licensed pilot by writing the Airmen's Certification Branch, AVN-460, FAA, P.O. Box 25082, Oklahoma City, OK 73125. Provide subject's full name and, if possible, the date of birth and/or SSN.

## 4.27 Locating the Homeless

The Salvation Army operates a locator service for finding homeless people. It will only help searchers who have a positive motive, e.g., those searching for lost family members. The Salvation Army will not tell you where the homeless person is, but will pass on the message that you wish to contact them. The regional addresses for the Salvation Army Missing Persons Service are:

- Eastern U.S.: 120 W. 14 Street, New York, NY 10011

- Central U.S.: 860 N. Dearborn Street, Chicago, IL 60610

- Southern U.S.: 1424 N.E. Expressway, Atlanta, GA 30329

- Western U.S.: 30840 Hawthorne Boulevard, Rancho Palos Verdes, CA 90274

The sharp rise in the number of homeless since the late 1970s has led to a plethora of private and local government agencies providing health care and counseling as well as operating soup kitchens and shelters. To a great extent these agencies have taken over the Salvation Army's traditional role. To find the appropriate agencies in your locality, look in the local community resources directory at your public library. In New York City, you can look in *The Source Book*, the *Directory of Community Services*, or the *Directory of Alcoholism Resources and Services*. The social service agencies listed in such directories are your most practical means of communicating with the homeless population if the Salvation Army can't help.

## 4.28 Government Locator Services

The Social Security Administration has one of the largest government databases of names and addresses. You can try reaching your subject through the SSA Letter Forwarding Service, Office of Central Records Operations, 300 North Greene Street, Baltimore, MD 21201. To use this service, you must give a humanitarian reason, such as a search for a lost family member, or else must be attempting to notify the person that money is due them (as from an inheritance). The SSA will not provide the address to you, but will forward a letter to the person you are seeking c/o the employer who filed the last quarterly earnings report for them, or c/o the address at which they are receiving Social Security benefits. You should provide the locator office with subject's full name, date of birth, and any other identifying information you have. The letter you wish forwarded must be sent to the SSA unsealed but can include a return address.

The IRS will also forward a letter to someone if you have a compelling humanitarian reason. You must provide the SSN, which is also the person's taxpayer I.D. number.

The Department of Veterans Affairs will forward a letter to any of the five million veterans listed in its files, i.e., any veteran who has ever applied for VA benefits. Call the VA regional office for your state to find out where to send your locator request. Then send an envelope to that address enclosing: (a) an unsealed, stamped envelope *without* your return address and with subject's name and VA file number or SSN on it; and (b) a letter to the VA requesting that the letter be forwarded and providing any information you have that would help in locating subject's records (e.g., full name, date or year of birth, approximate dates of military service, etc.).

You can also have a letter forwarded to a retired federal civil servant. Write: Office of Personnel Management, P.O. Box 45, Boyers, PA 16017.

The Federal Bureau of Prisons will locate a federal prison inmate for you; call (202) 307-3126.

## 4.29 Locator Form Letters and Mailings

If you are an adoptee searching for a birth parent, or if you are searching for a long-lost relative, try sending a form letter to a broad array of government agencies, private organizations, and periodicals that might have the person in their files and/or on membership, mailing, or subscription lists. Explain in the letter your humanitarian need to reach subject and provide any identifying information you have, such as full name, physical description, date and place of birth, SSN, mother's maiden name, spouse's name, and previous known addresses. State that you have enclosed a stamped envelope with a personal letter to subject inside it, and ask the organization or agency to either forward the letter or otherwise contact subject on your behalf. Urge the organization or agency to write you or call you collect if they have any questions about your request.

Make plenty of photocopies both of this letter and of a handwritten appeal to the person you are trying to find, and buy stamps and envelopes. Go to the public library and sit down with the *Encyclopedia of Associations, Washington Information Directory, Standard Periodical Directory,* and other relevant volumes. Address your mailing to any and every organization, agency or periodical that you think might have a past or present address for subject. Be sure to send the form letter to the circulation departments of the highest circulation general interest magazines as well as to specialty publications that might attract subject's interest (e.g., if your lost brother is a biker, you obviously would send the letter to magazines aimed at motorcyclists).

Except with the official government locator services (see above), it's difficult to predict how any of these third parties will react. They may have a fixed policy regarding such appeals, or it may be up to the person who opens the letter (or his or her supervisor) to decide on the spot. Even if the organization usually ignores such requests, your letter might touch the heart strings of the person opening the letter. Obviously a lot will depend on the care with which you word your appeal.

Unless told not to (as by the Department of Veterans Affairs), always include your return address on the envelope of the letter to be forwarded. If the third party forwards it to the address in its files, and the Post Office returns it to you marked "moved," you are one leg up: You now have a prior address for subject and can use the crisscross directory and other resources to eventually find a more recent address for subject or possibly the current one.

## 4.30 If You're Completely Stumped

Consider the following:

- Subject may be in an institution. According to the 1980 census, about 2.4 million Americans (more than 1 in 100) is in a prison, old-age home, long-term care facility, juvenile detention center, mental hospital, etc.

- Subject may be living overseas.

- Subject may be dead (see the Social Security Death Index).

- Subject may be living under a false identity, using either stolen ID or ID acquired in the name of a long-dead person.

- Subject may be living in a rural commune or some kind of totalitarian cult, mostly cut off from the rest of the world.

# 5.

## Backgrounding the Individual—Biographical Reference Works

### 5.1 Overview

Millions of Americans are included in one or more "who's who" type biographical dictionary or other reference work containing biographical data. There are thousands of such works: Never assume a person is not listed, no matter how humdrum his or her life might be.

Biographical dictionaries generally contain one-paragraph entries with noncontroversial basic information such as date and place of birth, names of parents, schooling, military service, job history, awards and other honors, names of spouses and children, membership in professional or fraternal societies, published writings, hobbies, and current office and/or home address.

Such information provides a springboard for further investigation. The date of birth may help you in ordering driver's license abstracts and other public records concerning subject. The name of a parent or former spouse may lead you to court papers regarding a divorce or probate of a will. Information about subject's educational background may guide you to college yearbooks, a master's thesis, or a doctoral dissertation.

As the biographical dictionary information on living persons is usually provided by the subjects themselves, some of it—even in the prestigious *Who's Who in America*—may be false or misleading. Indeed, since a subject will rarely wish to advertise his or her warts, most "who's who" type profiles will be slanted towards the positive. The late Teamster leader Jimmy Hoffa never described himself in a biographical dictionary questionnaire as a "labor racketeer." Nor did he ever list, under awards and honors, his multiple indictments in federal court. Nevertheless, most subjects tell the truth about noncontroversial basic facts such as when and where they were born, and this provides leads not only for further investigation but also a chronology around which you can organize your research findings.

## 5.2 Biography and Genealogy Master Index

Always begin with Gale's *Biography and Genealogy Master Index* (*BGMI*), which indexes almost 8 million biographical sketches from almost 1,800 editions and volumes of 635 source publications. Names of over 3.5 million living and deceased individuals are in this index. It not only lists each biographical work in which there is an entry for your subject, but all editions in which the entry appears. Thus, you can trace the changes in subject's entry from year to year, and glean from old editions information excised from the current one, e.g., former addresses and the names of former spouses.

The core of *BGMI* is an eight-volume base set published in 1980. Since then, annual updates have been cumulated every five years into subsidiary sets (1981-85 and 1986-90). There is also an abridged version covering 115 of the most widely available source publications. (Since the full version is easy to find in libraries and is also available on DIALOG, you should not restrict your search to the abridged version.)

Some of the *BGMI* source publications may be hard to find, especially the cited back editions (smaller libraries often discard them once the new edition is cataloged). If a publication is not available at your local library, get the research librarian to order the volume (or a photocopy of the cited entry) through interlibrary loan.

## 5.3 Marquis' Index and Search Service

The annual one-volume *Index to Marquis Who's Who Publications* (first published in 1974) is available like *BGMI* in most research libraries. It will direct you to biographies of over 280,000 persons in the current editions of eleven regional and professional Marquis Who's Who publications as well as *Who's Who in America*. The Marquis publications are also covered by *BGMI*, which provides (unlike the Marquis *Index*) cumulative indexing of past editions. The only advantage to the Marquis *Index* is that it provides indexing of current Marquis editions several months earlier than *BGMI* does.

Marquis offers a same-day search service that may be useful for paralegal researchers who have to gather background information fast for an upcoming deposition. If you cannot find your deponent in the Marquis *Index* or *BGMI*, call Marquis at (800) 621-9669. For a $35 fee, staff researchers will search the Marquis databases, which are updated daily, and fax you what they find (no charge if they fail to find anything). The researchers draw on Marquis' 2,500 volume collection of biographical dictionaries as well as its files of professional resumes and other research materials compiled for upcoming biographical sketches. Although they will not send you copies of materials from other publishers' dictionaries, they can provide you with information from various sources as they add it to the Marquis databases, which currently contain information on more than 500,000 Americans.

## 5.4 Biography Index; Bibliographical and Cataloging Works

*BGMI* only indexes biographical dictionaries. For other types of books and for periodical articles of a biographical nature, you should first consult the H.W. Wilson Company's *Biography Index*, a cumulative work going back to 1946. *Biography Index* issues quarterly indexes, interim annual cumulations, and permanent two-year cumulations. It covers every biographical-type article (including interviews and obituaries) in about 2,700 periodicals, together with more than 1,800 works annually of individual and collective biography.

*Biography Index* tells you if an article or book contains a photo of your subject. It also includes cross-indexing by profession or occupation (I usually do a quick scan to see if any of subject's colleagues or business associates are listed). Many public libraries have *Biography Index* from mid-1984 to the present in compact disk format. It is also available online from CompuServe.

Your search of *Biography Index* can be supplemented by a search of *Names in the News*, an index of biographical profiles, interviews, and obituaries from newspapers in over 450 cities. Cumulated monthly, quarterly, and annually, this NewsBank reference work is available in bound annual volumes and on CD-ROM. It dates back to 1978 and indexes by personal name about 20,000 articles a year. It includes many people who are only well known in their own localities (e.g., civic leaders, mayors, high school All-Americans) or who have unusual jobs that attract newspaper feature writers (e.g., Hollywood stuntpersons). The full texts of the indexed articles are included in the NewsBank Library, a microfiche resource available at many research libraries. (For more on NewsBank, see 6.4.)

Books of biography, autobiography, and collective biography, as well as published letters, diaries, and journals, are listed in R.R. Bowker's *Biographical Books 1876-1949* and *Biographical Books 1950-1980* (both are based on Library of Congress cataloging records). Bowker also offers the *International Bibliography of Biography*, 1970-1987, which lists over 100,000 biographies and autobiographies. For books of biography published since 1987, see Bowker's *Books in Print* titles. (How to search for books on various topics is covered in more detail in 11.3 and 13.2.)

If there is no biography on your subject, look for biographies of prominent people he or she has been associated with and check if subject is in the index.

## 5.5 Biographical Material Not Listed in the Master Indexes

As vast as *BGMI*'s range is, it skips over many important biographical reference works. Indeed, there is a vast unindexed universe beyond *BGMI*. Start with Robert B. Slocum's *Biographical Dictionaries and Related Works*. This is a standard guide describing 16,000 source publications. Even if you find your subject in *BGMI*, you might check in Slocum's (based on what you know about subject's background) to see if there are other books in which he or she is likely to be included. Slocum's describes many obscure local who's who's (many of them from decades ago—but they might include subject's parents), society registers, vanity registers, state government handbooks, works of collective biography, genealogical works, and bio-bibliographies.

Among the thousands of titles that can be used to supplement the standard biographical dictionaries, certain types stand out as being especially useful for journalists and private investigators:

- Society registers provide a means of tracing the marriages, divorces, re-marriages, yacht club memberships, and Ivy League academic credentials of America's upper crust. Preeminent is *The Social Register*, which, together with its supplement the *Social Register Summer*, reports annually on the most prominent families nationwide. *The Social List of Washington* (the so-called Green Book) covers society and officialdom in our nation's capital. Social registers have also been published for various states over the years (e.g., *The Social Record of Virginia*).

- State government handbooks (often called "blue books" or "red books") may contain detailed biographical sketches of state legislators and officials. Back editions and current editions of these handbooks nationwide contain upwards of 50,000 sketches of living Americans who are now or have been in state government.

- College class anniversary directories (usually published on the 25th anniversary) may include biographical sketches provided by the class members themselves. These sketches can be quite elaborate, providing data available from no other published source. To find out if your subject's class has published such a directory, call the college alumni association. The volume will most likely be in the college archives.

- "Vanity" directories solicit biographical information from ordinary people and charge them a flat fee to be included (the fee is often disguised as the advance purchase price of a deluxe, gold-embossed copy). At any given moment, there are about 200 vanity directory publishers soliciting from the American public. Typically, they will publish one or two editions of a work before replacing it with a new title. Thousands of such works have been published in the United States over the past century. They are of interest to an investigator for two reasons: first, they include people who would never get into a legitimate directory on their own merits; and second, they often devote a relatively long entry to a local notable who would only rate the briefest of entries in a legitimate work. Unfortunately, research libraries rarely buy vanity directories. You best hope is that the proud listee has donated a copy to his or her local public library or the library of the college from which he or she flunked out in 1948. If you find a copy, the self-written entry on your subject may include the fact that he or she previously appeared in another vanity directory (as if this were an honor of the highest distinction). By going from one to the other, you may notice odd variations in subject's life story that warrant further probing. The largest collection of vanity directories is at the Library of Congress.

- Local history buffs may have produced a biographical directory for your town, county, or state. Check with your county or state historical society.

- Family name books sometimes contain biographical dictionary-style entries on various living family members as well as their antecedents (e.g., the Doe family which emigrated from England in 1801). These books, also called genealogy books, are often self-published in very small editions by a family genealogy enthusiast for circulation mostly among his or her relatives or the members of a particular family name association. Many family name books are listed in the Mormon's Family History Library Catalog (see 4.11). Others can be found in *Genealogies in the Library of Congress* (look in the supplements as well as the base volumes) and *Complement to Genealogies in the Library of Congress* (the latter lists volumes that are *not* in the Library of Congress). In searching these reference works, be sure to look under John Doe's mother's maiden name as well as under the Doe surname.

    To find the right family name volume (if the surname is a common one), you may need the help of an expert. Consult the *Directory of Family Research* and also the *American Family Records Association— Member Directory and Ancestral Surname Registry*. These volumes match the names of hundreds of genealogists and family historians with the thousands of surnames they have researched.

    Once you find the right family name volume, it may turn out to have only genealogical lists and charts. This of course is useful information, but you were hoping for a collection of biographical sketches. In such cases, contact the author—he or she may have files containing very detailed biographical data. (If the volume includes biographical sketches, I would still contact the author to see if he or she has additional data.) You might also contact other members of the relevant family name association.

Although your chances of finding a biographical sketch of your subject in a family name book are not great, it's worth a try. You could end up with a wealth of material about subject and his or her children, parents, siblings, grandparents, aunts, uncles, and cousins. If subject still lives in his or her hometown, surrounded by these relatives, information about the extended family could be quite important in researching subject's business affairs.

- A vast amount of biographical material is contained in professional, organizational, or alumni directories or rosters. The *BGMI* does not index them (nor does Slocum's list them) because the amount of biographical detail per entry is just too sketchy. Yet if you can find from such a directory where a person lives and works (and from back issues, where they used to live and work), their year and place of birth, their college and year of graduation, and their spouse's name, you have made a good beginning. There are a vast number of such directories—thousands are published each year by college alumni associations alone. I have found that although each volume listing your subject may only contain one or two facts not included in the others, the amount of information builds up when you go through volume after volume. One individual I was tracking was listed in two alumni directories, two national faculty directories, a law directory, and a directory of consultants. By the time I finished looking through these, I had almost as much material as from a brief *Who's Who* entry. Such directories can also be useful in finding sources who will tell you their recollections of your subject—professional colleagues, fellow faculty members, former classmates, etc. In addition, some professional directories may contain fullblown biographical sketches of selected persons: For instance, *Martindale-Hubbell* includes sketches on about 40,000 attorneys along with its roster of over 800,000.

In searching a professional directory or roster, don't just look at the entry for your subject; the volume may contain other gleanings as well. For instance, take the annual directory of the Special Libraries Association, a national organization. As well as having an alphabetical listing of member librarians and their work addresses, it has a list of members by city or state chapter and a list by subject specialty division, both of which could help you identify members who might be close colleagues of the librarian you are backgrounding. Furthermore, the directory includes lists of SLA charter members, honorary members, Professional Award and Special Achievement Award holders, Special Citation recipients, SLA Hall of Fame members, SLA past presidents, current officers (both nationally and for each chapter and division), SLA committee members and officers, SLA representatives to other professional organizations, and a name index with page citations for each listing of each member's name. On top of this—should you wish to engage in personal surveillance of your subject librarian—the directory even gives you a list of twelve upcoming meetings and the city and hotel at which each will take place. (It also gives you subscription information on the SLA's monthly newsletter, *SpeciaList*, which might have much more detailed information about subject's career.)

For more on professional and other directories and rosters, see 4.15.

## 5.6 Back Editions of Biographical Dictionaries

If you have found an entry on subject in the current edition of a biographical dictionary, don't neglect the entries for past years in this and other dictionaries. As noted above, the earlier sketches may provide you with past residential and business addresses and the names of former spouses that are not included in the most recent entry. By going to these earlier sketches in the library stacks, you may be able (if subject is consistently included) to use the back editions almost like a city directory.

Although *BGMI*'s indexing of back editions dates back only to 1974-75, a few standard reference works have their own cumulative indexes covering earlier years. For example, *American Men and Women of Science*'s index to its first fourteen editions (1901 to 1979) includes over 270,000 living and dead scientists. In searching most biographical dictionaries prior to the mid-1970s, however, you will have to check each edition separately in the library stacks.

Often a subject who is not included in any current biographical reference work was included years ago because of a government appointment, electoral candidacy, or some other factor that temporarily qualified him. To find sketches not indexed in *BGMI*, look in *Directories in Print* and Slocum's for works in which his name might have appeared given what you know about his past. For instance, let's say you are backgrounding John Doe, a local public relations consultant. You know from the rather meager press clippings that he was a Congressional staff aide for a brief period in the early 1970s for the late Congressman Mark Grouch. Looking in *Directories in Print*, you see a listing for the annual *Congressional Staff Directory*. Finding it on a nearby shelf, you note that it includes in its current issue 3,200 biographical sketches, many of them rather detailed. You also note that it has been published since 1959. If you can get the back editions for the early 1970s from the library stacks, you will probably find a sketch of Mr. Doe.

## 5.7 Parallel Backgrounding Using Biographical Reference Works

Information that your subject has failed to provide to the compilers of a biographical work may be contained in a sketch of one of his or her relatives, business partners, or close friends. In an investigation of a local attorney, I found biographical sketches that contained virtually nothing about what I was interested in—his connections to the Arab world. Then, I found a sketch in a biographical dictionary about his closest friend from college days, who was described as a lobbyist for Arab governments and a partner in a Mideast trading firm, the name of which included subject's surname. I was later able to confirm that subject was indeed connected to this firm.

To utilize parallel backgrounding to the maximum I suggest you keep a list of names of your subject's associates and relatives as you find them. Periodically check the biographical indexes again to see if they include any of the latest names you've collected. Also, as you search the indexes develop a roster of persons who have the same surname as your subject or as a married woman subject (or a male subject's wife) prior to marriage. When you check any biographical dictionary that has an entry for your subject, alway look at the surrounding same-surname entries. And if your subject comes from a successful, highly educated family, always check the same-surname entries in *Who's Who in America*.

Parallel backgrounding works best when you find what I call a "high-yield" biographical dictionary—one that specializes in a category of people likely to include a high percentage of subject's cronies. Once I was looking into the background of a Harlem businessman. The world of black New York politicians and businesspeople is unusually tight-knit—everyone has dealings with everyone else, and since they live in the city that has long been the cultural center of black America, a large percentage of them are included in *Who's Who Among Black Americans*. I found myself returning to this book again and again.

If I were backgrounding a prominent New York corporate attorney, *Who's Who in American Law* and *Who's Who in America* would be high-yield books sure to include many of his or her colleagues, friends, and clients. In addition, if the attorney was from an old-money family I might use *The Social Register* to ferret out the more subtle interrelations.

## 5.8 The Clues Hidden in Biographical Sketches

The fact that the information for biographical sketches is provided by the sketchees themselves (in effect, is subject to censorship) can be turned to your advantage. Study the sketches for clues inherent in the facts as presented. For instance: Are there any glaring or subtle contradictions between the versions of subject's life that he or she provided to different directories (or to the same directory) at different times? Are there any unexplained time gaps? Does subject list himself or herself as a board member of a corporation, bank, or foundation that you can't find in any business directory? Does he or she claim a degree from a college that is not listed in the directories of accredited institutions? Does he or she list attendance at a college without claiming a degree—if so, what happened? Does he or she claim personal achievements that sound dubious on the face of it (e.g., an alleged Rhodes scholarship for someone whose life before and after does not fit the pattern)? (Note that some legitimate directories try to verify such claims; others let them pass unless they are embarrassingly obvious.)

Politicians occasionally get caught listing exaggerated or false information in their campaign bios. You may find that your subject has done likewise in his biographical dictionary entries (which is one reason you should develop an efficient filing system for all the background information you collect).

## 5.9 Biography Database Searches

Biographical dictionaries currently available online from DIALOG include *Who's Who in America* (77,000 records of current top professionals in many fields); *American Men and Women of Science* (more than 125,000 profiles of active scientists and engineers); *Standard & Poor's Register—Biographical* (over 68,000 profiles of top executives); *Who's Who in American Politics* (24,000 sketches); and *Who's Who in American Art* (over 11,000 sketches). ORBIT Search Service offers the online version of *Who's Who in Technology* (almost 38,000 entries). The NEXIS People Library offers the *Almanac of American Politics*, Associated Press Candidate Biographies, and Congress Member Profiles, as well as biographical stories from the *New York Times*, *Washington Post*, and *People* magazine. For other biographical databases that may come online soon, see the subject indexes in *Computer-Readable Databases* and *Information Industry Directory*. Look not only under "Biography" but also under the heading for your subject's profession.

Although the number of biographical works online is still rather skimpy, you can compensate via full-text cross-searching. *Who's Who in America* is especially useful in this respect: Prominent people tend to associate with other prominent people, and thus any person listed in *Who's Who in America* is almost guaranteed to have ties to other listees. Do you need sources for an article about Mr. Doe? Search for listees who are the same age as Doe and come from the same hometown, or who attended the same university as Doe during the same years, or who worked at Corporation Y while Doe was an executive there, or who today live in the same city as Doe and belong to some of the same professional or fraternal organizations, or who married women with the same maiden name as Doe's wife (this may help you find Doe's brother-in-law). If you are a reporter doing a friendly feature story on Doe, probably the majority of these people will be willing to talk to you. If you are an investigative reporter, your search may turn up Doe's bitterest enemy.

To supplement online searches, try the CD-ROM databases at your local research library. Biographical directories and dictionaries available on CD-ROM (but not yet online) include *Martindale-Hubbell* and portions of *Who's Who in Finance and Industry* (the latter, as part of LOTUS's CD/Corporate Database). Marquis' *Directory of Medical Specialists* (365,000 names) will become available on CD-ROM in 1992. Look for many more biographical reference works on CD-ROM in the next few years; they will be listed in *CD-ROMs in Print*.

# 6 ∙

# Backgrounding the Individual—Newspaper and Periodical Searches

## 6.1 Overview

If your target is a celebrity or an elected public official, newspapers and periodicals are the obvious place to start your search. But newspapers, especially local dailies, can also provide a wealth of easily accessible data about tens of millions of noncelebrities, from stockbrokers to muggers and from debutantes to shopping bag ladies. The back issues of America's newspapers and periodicals—and the knowledge and working files of the vast number of journalists working for these publications and for wire services—are a potential intelligence resource to rival the combined assets of the CIA and KGB.

The searching of newspapers and periodicals is done easiest by computer. An ever-increasing number of publications are available online. You can search the entire contents of every issue of a newspaper for, say, the last five years, finding every mention of subject and his or her associates. Indeed, you can do full-text global searches of hundreds of publications at once via large database networks such as NEXIS.

The vast majority of newspapers and periodicals, however, are not yet included in the electronic information net, and of those included most are only available online for issues dating back less than ten years. If you need to search a relatively obscure publication that is not online or an issue of an online publication that precedes the publisher's adoption of an electronic news storage system, you will have to use traditional low-tech resources. These include printed or microform indexes and abstracts (although some of these are online for the more recent years), newspaper "morgues," archival clippings files, and microform or bound volume back-files. The online databases, however, may show you where to look. An online article will often refer to a pre-online incident that the reporter read about in old newspaper clips while preparing his or her own article. Usually the reporter will give the date or at least the year of the prior incident/article; you can then find the article on microfilm, and perhaps learn from it about a still earlier incident/article.

## 6.2 Serial Directories and Catalogs

In backgrounding a prominent Midwest businessman, you may decide to search the daily newspapers in his state as well as local and national business and trade publications. In a once-over investigation, online databases may give you everything you need. But what if you want to search smaller dailies (and weeklies) in subject's locality that are not online? Or pre-online issues of the major local daily? Or the issues of its now defunct rival? And where can you find these publications on microform? The following directories and catalogs are available in most research libraries:

- *Gale Directory of Publications and Broadcast Media* (formerly *Ayer Directory of Publications*). This three-volume set enables you to quickly identify by city and state/province about 36,000 newspapers and periodicals in the United States and Canada. It includes every type of publication except newsletters, house organs, publications issued less than four times a year, and publications issued by primary and secondary schools or houses of worship. Defunct publications are removed from the main body of entries and listed as "ceased" in the master name and keyword index. (For further listings in a given city, see the local yellow pages.)

- *Standard Periodical Directory*. Includes annotated entries on 65,000 periodicals in the United States and Canada, divided into 230 major subjects and featuring a title index and cross-index.

- *Newsletters in Print*. Includes over 10,000 newsletters arranged under thirty-three subject headings.

- *Newspapers in Microform*. This cumulative Library of Congress reference set (with annuals between cumulations) includes religious, collegiate, labor, and other special-interest papers, as well as general news dailies and weeklies. It also covers defunct and merged papers. Under the current title of a paper will be listed its previous names and the names of papers that merged with it or split from it, with the dates of each change, e.g., if you look under the *Anytown Courier-Herald* you will find that it is the successor to the *Anytown Courier* and the

*Mist County Herald,* which merged in 1949. *Newspapers in Microform* tells which libraries, archives, or microform companies at home or overseas have copies of the given newspaper; it also indicates partial or badly broken runs.

- The *Union List of Serials* and *New Serial Titles.* These multi-volume sets, available in major research libraries, include many obscure and long-ceased serials that are not listed anywhere else. They will tell you which libraries in your region have a particular serial.

- University Microfilms International's annual catalog of newspaper microform backfiles. The 1990 catalog lists over 7,000 national and international newspapers and the years covered. Any backfiles in this catalog will probably have been purchased by the major public or university libraries of the region in which the paper is or was published.

## 6.3 Full-Text Searching

Database searches are conducted via key words or phrases supplied by you: for instance, the full name of the person you are researching, or the names of businesses he or she is associated with, or the names of his or her closest cronies. The database vendor's computer will search through every word of every text in the entire database or any portion thereof and transmit to your computer—for screen display or a printout—the title, date of publication, and source of each article in which the key word or phrase appears. You can then decide which texts you wish to print out.

The most obvious advantage of newspaper database searching is that you avoid the tedium of going through indexes and microfilm. Equally important, database searching enables you to find articles in which your subject is only mentioned in passing—those obscure, unindexable references that so often furnish a researcher with the best leads but are only found by sheer luck when one is scanning the headlines on microfilm.

Full-text-search capabilities also facilitate research into your subject's associates and business interests. Let's say you are compiling information about Moe G., a Midwest racketeer. Your search of the database that includes the leading daily in Moe's city (online since 1981) turns up six articles, from which you learn that Moe owns a trucking company and a nightclub, and that he is alleged to control a Teamster local. The articles also tell you the name of the lawyer who won Moe's acquittal in a 1984 extortion trial and the names of Moe's co-defendants. You then search the database for every article mentioning Atlas Trucking, the Starlite Lounge, IBT Local 4294, or any of the individuals mentioned in the six articles. You find twelve additional articles via this second search. Some have no probable connection to Moe (i.e., an article about the death of Moe's lawyer's mother in a nursing home). Others spark your interest—

for instance, the article about the Atlas driver arrested for armed assault in 1982. You can take your search through additional cycles, if you believe the results will warrant the expense. (Understand that this example of a search restricted to one newspaper is somewhat simplified; in reality you would be doing a global search of other publications in the database along with the local daily. You might also enter another database from another vendor that carries the city's rival daily. And you would surely search a variety of nonnewspaper databases—for instance, the Dun & Bradstreet databases for Moe's trucking business and the computerized indexes of the local courts for all names connected to Moe.)

## 6.4 Newspaper Databases

### Online Full-Text-Search Newspaper Databases

The number of newspapers searchable via databases is constantly growing. As of early 1991, major database vendors were offering over 150 daily papers from every region of the United States and Canada. Gale's *Computer-Readable Databases* (*CRDB*) tells you which vendor offers which newspapers (look first at the master list of online papers in the subject index). If the one you need is not listed, it may have become available since the last edition of *CRDB* was prepared: Check with the newspaper's librarian.

Four database vendors dominate the market for full-text newspaper searches:

- Mead Data Central's NEXIS offers over 100 U.S. and foreign newspapers. These include the *New York Times* (since 1980), the *Washington Post* (since 1977), the *Los Angeles Times* (since 1985), most important regional U.S. newspapers, and business and law newspapers such as *American Banker, Computerworld, Legal Times,* and the *National Law Journal.*

- DataTimes provides full-text searches of almost ninety dailies in the United States and Canada, mostly beginning in the middle to late 1980s. Newspapers covered include the *Washington Post,* the *Washington Times,* and the *Wall Street Journal,* as well as local dailies such as the *Orange County (CA) Register* and the *Dallas Morning News.*

- VU/TEXT provides online access to the full text of seventy-two daily newspapers, ranging from the *Anchorage Daily News* to the *Miami Herald.*

- DIALOG—the online network most widely available at public libraries and on college campuses—offers a rapidly growing collection of newspapers (thirty-four as of 1991) via its "Papers" file.

All of the above databases except NEXIS can be accessed through CompuServe. You can subscribe to all directly, or gain access through your public library's computer search service.

### Full-Text-Search Newspaper Databases on Compact Disk

Many newspapers are making their backfiles available on CD-ROM for sale primarily to libraries (see *CD-ROM Periodical Index*). Although updated less frequently than their online counterparts, CD-ROM databases can be searched for free in your local public library. Among the many newspapers that have announced publication of back issues on CD-ROM are the *New York Times, Wall Street Journal, Washington Post,* and *Los Angeles Times.* Newsbank, Inc. is currently offering the full text of the *Arizona Republic & Phoenix Gazette, Boston Globe, Chicago Tribune, Christian Science Monitor, Dallas Morning News, Ft. Lauderdale Sun Sentinel, Orange County Register, Orlando Sentinel, The Oregonian, Sacramento Bee, San Francisco Chronicle, St. Petersburg Times,* and *Washington Times.*

### The "Hidden" Newspaper Databases

Many of the online newspaper databases offer full-text coverage only back to the middle or late 1980s; very few provide coverage for issues prior to 1981. However, a given newspaper might have earlier issues stored electronically which have not been offered for sale through an online vendor because the demand is not sufficient. Also, some newspapers have had electronic storage for years but have not yet made a deal with a vendor to sell even their most recent issues.

If you need access to any of these "hidden" databases, call the newspaper's library. They may occasionally do courtesy searches for serious researchers. Some publications have a reader call-up service, linked to the inhouse database, through which you can at least get the date of the article you need.

You can also try contacting the staff reporter whose beat corresponds most closely to what you are working on, get him or her interested in your research, and obtain printouts from the electronic library—and photocopies from the clippings morgue that predates the electronic library—in return for news tips or a promise of access to the fruits of your investigation.

### Full-Text-Search Newswire Databases

Most people think of the wire services as organizations with correspondents in Washington and Moscow who cover the Big Picture. But the wire services' state and regional bureaus, and their stringer correspondents, generate a vast amount of news at the grassroots. United Press International has full-time staff in every major U.S. metropolitan area, as well as overseas, and a network of over 3,000 stringers. Associated Press employs more than 1,000 journalists in 141 domestic and 83 overseas news bureaus; as a news cooperative, it also draws on the resources of its 1,300 newspaper members and 5,700 radio-television members.

Wire service databases can be accessed through NEXIS, which includes AP and UPI's world, national, business, and sports wires (since 1977 and 1980 respectively), UPI's state and regional wires (since 1980), Southwest Newswire (since 1984), Central News Agency (since 1984), and Gannett News Service (since 1989). NEXIS also offers foreign wire services such as Reuters and several publicity, business, and financial newswires.

## Online Newspaper Abstracts and Indexes

The Information Bank, produced by the New York Times Company and available through NEXIS, offers abstracts of *New York Times* articles beginning in 1969 (as well as the full text of the *Times* from 1980 on). It also selectively abstracts about thirty other national and regional newspapers including the *Wall Street Journal, Washington Post, Atlanta Constitution, Chicago Tribune, Houston Chronicle, Los Angeles Times, Miami Herald, San Francisco Chronicle,* and *Seattle Times,* as well as scores of periodicals.

University Microfilms International (UMI) indexes and abstracts twenty-five national and regional newspapers. Coverage since Jan. 1, 1989 is included in UMI's Newspaper and Periodicals Abstracts (formerly Courier Plus) database which also includes over 1,000 periodicals. For 1984-88 coverage, see UMI's Newspaper Abstracts database. Both are available through DIALOG. (UMI also offers CD-ROM coverage of nine newspapers back to 1989 via Newspaper Abstracts Ondisc.)

The National Newspaper Index, available through DIALOG, provides front-to-back indexing of the *New York Times, Christian Science Monitor,* and *Wall Street Journal* from 1979 on, and selective indexing of the *Los Angeles Times* and the *Washington Post* from 1982 on.

## NewsBank Electronic Information System and Newsbank Library

The NewsBank Electronic Information System provides selective indexing of newspapers in over 450 U.S. cities, including the state capital and largest city of each state (for NewsBank's coverage of periodicals, see 6.5 below). The system is offered on CD-ROM back to 1981 with microfiche coverage back to 1972. All articles included are locally written—the index does not include wire service or syndicated articles. Indexing topics are divided into sixteen categories of current interest and a "Names in the News" category; under these are hierarchically arranged subtopics so you can quickly narrow your search. Issued concurrently with the index is the NewsBank Library which offers the full-text on microfiche of over 2 million of the indexed articles back to 1972 (this resource will probably become available on optical disk in 1992, but with coverage only back to 1990).

This superb research tool can extend the range of your search far beyond the current limits of online databases. The system is available at over 5,000 libraries nationwide.

## 6.5 Periodicals Databases

### Online Full-Text-Search Periodicals Databases
NEXIS offers over 660 general-interest, business, and trade periodicals online. Most of these can be searched globally via the NEXIS Magazine Files. You can also search periodicals, newspapers, and wire service releases all together in specialized group files, such as Business, Finance, and Trade/Technology, or in the various regional files. The NEXIS Newsletter Files include over 140 newsletters, mostly in business and technology. For more comprehensive coverage of newsletters, see NewsNet's databases, which offer the full text of over 430 newsletters and other publications in dozens of subject categories. Also see Magazine ASAP (available from DIALOG), which allows full-text searches of over 120 general-interest magazines from 1983 to the present.

### Online Periodicals Abstracts and Indexes
Over 1,000 general interest, professional, and scholarly periodicals are indexed and abstracted in UMI's Newspaper and Periodicals Abstracts database back to 1988 (with selected titles back to 1986). UMI's ABI/Inform offers abstracts/indexing of about 800 business-oriented titles with coverage of some back to 1971. Both of these databases are available via DIALOG. Note that there is an overlap of about fifty titles, so if you need a pre-1988 search of a Newspaper and Periodicals Abstracts title always check ABI/Inform.

Magazine Index, an Information Access Company product available through DIALOG, covers about 500 general-interest magazines, with some records dating back to 1959. (See also Magazine ASAP above.)

Numerous other indexes and abstracts are available through DIALOG and other database vendors (see 11.3 for scholarly and scientific periodicals and 12.2 for business periodicals). If the index in question does not include personal names, a search of appropriate subject listings may turn up articles in which the person you are backgrounding is mentioned.

### CD-ROM Periodicals Indexes and Full-Text Retrieval
UMI's Periodical Abstracts Ondisc is an index/abstract to 450 general periodical titles dating back as far as 1986; through UMI's General Periodicals Ondisc, you can get electronic page images cover-to-cover of about 200 of the indexed titles back to 1988. UMI also offers ABI/Inform Ondisc, about 800 business titles (some records back to 1971): You can get page images of about 300 of these titles back to 1988 via Business Periodicals Ondisc.

The NewsBank Electronic Index to Periodicals, available in many public libraries, is a comprehensive CD-ROM index of 100 general-interest magazines beginning with the January 1988 issue. Cover-to-cover microfiche editions of selected periodicals are also offered.

A crucial CD-ROM and diskette resource for investigative journalists is NameBASE, an index of personal, corporate, and organizational names that have appeared in magazine articles dealing with national security, the CIA, narcotics trafficking, and political extremism. Also entered into NameBASE are the indexes from over 350 books. This database, which includes over 100,000 name citations, is available from Public Information Research in Arlington, Virginia.

For more information on CD-ROM products, see the CD-ROM *Periodical Index*.

## 6.6 Full-Text Copy Services and Interlibrary Loan

If the index/abstract you are using does not offer direct microform or optical disk retrieval of the cited article, you may find it in bound volumes of the publication or on microform from another publisher at your public library. If not, you can obtain the microform or the bound volume (or a photocopy of the particular article) through interlibrary loan.

Another option—if you need the article quick—is to go through a commercial copy service. You can order virtually any article indexed by DIALOG through an online service called DIALORDER. DIALORDER will route you to the appropriate company (one of about eighty full-text copy services) for copies of articles from the periodical or newspaper in question. This company will send you the copy by mail or fax and bill you separately.

If you are using a non-DIALOG index, contact the vendor or publisher regarding such services (hardcopy indexes may include a copy service telephone number in the front of the book). Note that University Microfilms International will provide copies, through its Articles Clearinghouse, of articles indexed in UMI databases. UMI's indexes cover over 1,000 publications, often back to 1971. If you find an article in a non-UMI index, it is often in UMI's as well. The Articles Clearinghouse is thus a service to keep in mind. The number is (800) 521-0600, Ex. 2533 or 2534.

## 6.7 Using Database Searches to Identify Relevant Hardcopy Files

An article found in NEXIS may be based on another article from a paper that is not included in any database or print index. But even with no citation, you can often figure out where to look. For example, if the *New*

*York Times* publishes a short article on the antics of a right-wing extremist group in Iowa, the *Times* stringer has probably just followed up on a much more detailed story or series in a local Iowa paper. To decide where to look, keep in mind the following:

- A major local or regional story, although covered in major newspapers in other regions, will usually be treated in greater detail in the newspapers of its own region.

- A local story will often be treated in greater detail in a small local daily or weekly than in a major metropolitan daily fifty miles away.

- A local daily may be taking its story from a rival local daily or from a local weekly that treated the story in greater detail.

- Newspapers rarely give proper credit to each other.

## 6.8 Print Indexes to Newspapers and Periodicals

### The New York Times Index

*The New York Times Index* is the grand old lady of newspaper indexes. Much larger and more detailed than any other index, it provides coverage back to 1851, and has been published in roughly its present index/abstract format since 1913. If your subject has been prominent in a part of the country distant from New York, don't automatically assume his or her name is not included. The *Times* is America's newspaper of record—truly national in scope—with bureaus or stringers in every region.

The *Times Index* is an annual, with monthly supplements cumulating quarterly. Unless you access it via database (or use the personal-name print index described below), you must search the volumes one by one. I find it chiefly useful in parallel and indirect backgrounding (see 1.3 and 1.4). Indeed, its hierarchical, topic-oriented mode of organization and its chronological abstracting of articles on each topic will give you a unique sense of your subject's interrelations with the people and institutions surrounding him or her. For instance, let's say you are backgrounding a Teamster official, and you want to know about his rise in the union in the 1950s. Irrespective of whether there is any mention of him in the *Times Index*, I would suggest you photocopy everything under the headings for Teamsters Union and Organized Crime for every year during that period. Take the material home for careful study—it is likely to contain a wealth of leads to unindexed articles, old court files, and subject's former associates or opponents. (As you learn more about subject, you should consult these photocopies again from time to time to see if you've missed anything.)

## Personal Name Index To The New York Times Index

*The New York Times Index* can be a tedious tool for searching out references to an individual (or his or her associates) over a stretch of years: Entries for individual names merely refer you to subject entries, and you thus have to look up a name at least twice within each annual volume. The problem of searching quickly for personal-name references without going online is solved in part by a remarkable reference work, the *Personal Name Index to the New York Times Index*. The base set is twenty-two volumes covering 1851 to 1974. A five-volume supplemental set brings things down to the present and also corrects errata and adds names missed in the base set. Both sets are organized alphabetically by personal name— you only have to look in two volumes (one for the base set, one for the supplemental set) to find every listing of your subject's name.

The *Personal Name Index* lists each reference to a person's name in chronological order, providing the year and page in the *Times Index*. To find out more you must go either to the *Times Index* volumes or (beginning in 1979) to the online National Newspaper Index. Then you must decide whether to access the full text either from the microfilm (back to the *Times'* beginning) or from the online abstracts (since 1969) or from the full text online (since 1980). If there are only a couple of references to your subject and you're in the library anyway, you'll probably just go to the microfilm. Although the *Personal Name Index* is NOT an index to the full text of the *Times*, skillful use of it can help you find articles in which unindexed references to subject (or mention of his or her activities without mention of his or her name) occurs. Simply look under the names of subject's closest associates, especially those who are better known than subject (or were better known during the years in question). For instance, I decide to background New York City Councilman X. I know that in the early 1960s he was an aide to Congressman Y. I do not find him in the *Personal Name Index* for those years, but I do find Congressman Y. Checking in the *Times Index*, I find that several of the articles on Congressman Y concern a bribery scandal. I then look at the text of the articles on microfilm and discover that the future Councilman X's name was mentioned several times in connection with the scandal. This stimulates me to try to learn more about this all-but-forgotten incident.

## Print Indexes To Other Daily Newspapers

University Microfilms International offers print indexes of the following newspapers: *American Banker* (since 1971); *Atlanta Journal & Constitution* (since 1982); *Boston Globe* (since 1983); *Chicago Sun-Times* (1979-82); *Chicago Tribune* (since 1972); *Christian Science Monitor* (since 1945); *Denver Post* (since 1976); *Detroit News* (since 1976); *Houston Post* (since 1976); *Los Angeles Times* (since 1972); *Nashville Banner & Tennessean* (since 1980—only on microfilm); *New York Times* (since 1851); *St. Louis Post-Dispatch* (since 1975); *San Francisco Chronicle* (since 1976); *Minneapolis Star & Tribune* (1984-85 only); *New Orleans*

*Times-Picayune* (since 1972); *USA Today* (since 1982); *Wall Street Journal* (since 1955); and *Washington Times* (since 1986). These indexes are not as detailed as the *New York Times Index*. If your subject is not listed, look under the subject headings with which his or her name is most likely to be linked, then search the most promising articles on microfilm.

For other indexes to daily papers, check Scarecrow Press' three-volume *Newspaper Indexes: A Location and Subject Guide for Researchers*. Many of the indexes it lists are only of interest to historians and genealogists, but it also includes indexes for some contemporary newspapers. Many of the listed indexes are not continuous, offering only spotty coverage. Some are unpublished in-house indexes or are listed as being in preparation. If you don't find a particular newspaper listed, don't assume there is no index: Always check with the newspaper's librarian. There may at least be a card file index of obituaries.

If your library does not have the microfilm backfiles of a newspaper whose index you have searched, you can obtain the microfilm reels through interlibrary loan or photocopies of specific articles via an article delivery service (see 6.6).

### Print and Microform Magazine Indexes

Magazine Index (see 6.5) is available on microform at many public libraries. Its coverage dates back to 1959. For earlier coverage, go to *Readers' Guide to Periodical Literature*, which has indexed the nation's most important weekly, monthly, and quarterly magazines since early in this century. *Readers' Guide*, today covering 174 periodicals, is an annual index with cumulative supplements between volumes. It is available online back to 1983; earlier volumes must be searched one at a time. The personal name indexing covers only the articles' authors and the persons who are the chief subjects of these articles (for the latter, you're better off using *Biography Index*).

The rather stodgy list of publications in *Readers' Guide* is supplemented by *Popular Periodicals Index* (about thirty-five magazines), *Access: The Supplementary Index to Periodicals* (about 120 magazines and weekly newspapers), and *The Left Index*. These indexes include a number of publications specializing in investigative and advocacy journalism.

### Print Indexes Of Business Newspapers and Periodicals

Described in 12.2, these indexes are rich in biographical data about individuals in all fields, not just business.

## 6.9 Morgues of Local Dailies

Morgues (systems of crossreferenced and crossfiled clippings files) are kept by most daily papers. In recent years many papers have turned to computer databases stored on optical disks to replace clippings files. Many of these databases are then leased to database vendors who in turn sell online access to the general public. So, in essence, when you perform a full-text search of a newspaper you are accessing its morgue. But these electronic morgues rarely date back more than ten years. For earlier material, reporters must rely on the old-fashioned clippings files. The best of them will contain every mention of your subject, his or her business firm, etc., going back many decades.

Even if there is an index to your local paper, access to the morgue is of great value. First, it will save you the tedium of going to microfilm at the public library. Second, a morgue that has a really thorough system for clipping the daily editions will be much more complete than an index. Third, the morgue may contain court papers and other documents gathered by reporters in the course of their investigations. Fourth, it may contain clippings from the local weekly, ethnic, or "underground" papers (which usually can't afford their own morgues) as well as from rival dailies in the city and surrounding region. Fifth, the newspaper may have obtained the clippings files of defunct local dailies and merged these files into its own.

Some newspapers will allow limited morgue access to scholars, freelance journalists working on books, or researchers from public interest organizations. If not, contact one of the newspaper's staff reporters or part-time stringers (or a freelancer who often writes for the paper's weekly magazine or Op-Ed page). Interest him or her in your investigation, and arrange an exchange of information, including morgue clippings.

## 6.10 If There's No Index or Morgue

Don't despair. A longtime reporter may remember an article on your subject and the approximate date. Or you may learn through one of subject's former neighbors or a biographical dictionary the approximate date of an event in subject's life that might have been reported (for instance, subject's marriage or the marriage of one of his or her children). In these cases, you will have to do some searching through the microfilm or bound copies, but at least you've narrowed your search within reason.

An event covered by an unindexed local paper may also have been covered by a larger, regional paper that does have an index. Learning the date of the event from the indexed paper, you then go to the microfilm or bound volumes of the unindexed paper.

Federal, state, and local court indexes will tell you about civil and criminal cases involving subject that may have been reported in the press. Court cases can drag on for years, but the most important news articles usually appear at predictable times. In your average low-profile criminal cases, this will be the next issue following the arrest and possibly the next issue following the jury's verdict or the sentencing. For high-profile cases, you will have to search through the dates covered by the trial, but look

especially for articles regarding the opening arguments and the summa-
tions. For civil cases, news coverage is most likely to come when the case
is first filed (especially if the plaintiff calls a press conference) and thereafter either when the jury announces its decision or the parties to the action
announce an out-of-court settlement. (To find these dates, examine the
docket sheet at the courthouse.)

If the above methods fail to turn up any articles, you might try a local
muckraker's clippings files (see 6.24).

## 6.11 Suburban News

Your subject may work in the city, but live in a suburban community (or
a smaller nearby city) that has its own daily newspaper. Suburban papers
are gradually becoming available online, e.g., *The Record* in Hackensack,
New Jersey. They are often the best source for news about subject's social
life, civic activities, and grassroots involvement in electoral politics.

An area's major metropolitan daily may produce special editions for
various suburbs to compete with the suburban-based dailies. Generally, a
major metropolitan paper's article on a suburban event which you find in
the paper's index or through a database search will have been treated at
greater length in the edition for that suburb than in the metropolitan
edition or in editions for other suburbs. Indeed, it may have appeared
ONLY in the given suburban edition. Although you will be able to find
the article in a properly organized morgue, you may not find it (or may
not find the full version) on the microfilm or in the database if these only
include a single edition.

## 6.12 Defunct and Merged Dailies

The importance of newspapers in the United States has declined steadily
since the advent of television. If you are backgrounding a prominent person of middle age in any medium-sized or large American city, there is a
good chance that at least one local daily newspaper—a newspaper that
might have reported on your subject's activities—has gone out of business
or merged with another since your subject became an adult. Indeed, a
local daily may have gone out of business or merged with another since
your subject's rise to prominence (say, within the last ten years). To find
the names of defunct and pre-merger papers, check *Newspapers in Microform* (see 6.2). For papers not in microform, consult the back issues of
*Editor & Publisher Yearbook, Ayer's* (now the *Gale Directory of Publications and Broadcast Media*) and/or the local yellow pages. The microfilm
or the bound volumes of a defunct newspaper may be available in the
public library of the city in which it was published, or in the library of a
surviving local daily, and there may be an index of some kind. In addition,

the defunct paper's morgue may have been sold to a surviving daily or donated to the local public library or county historical society.

## 6.13 Local Weeklies

In every metropolitan area, you will find flourishing weeklies, both of the free (shopper) variety and of the paid subscription/newsstand type, aimed either at the entire city or at a particular city neighborhood or suburban community. You will also find weeklies specializing in ethnic or alternate-lifestyle news.

A few weeklies will have excellent clippings files organized like those of the dailies. Most weeklies lack this, but may keep files on ongoing local political conflicts or the paper's most important investigative pieces through the years. In addition, the editor may remember an article on your subject and be willing to dig out the back issue in which it appeared.

Never underestimate small weeklies. Major investigative pieces by large dailies frequently are based upon spadework performed by the weeklies, and the latter may treat the subject in much greater detail (and with much less pulling of punches) than any large daily would.

*The Alternative Press Index* and *Access: The Supplementary Index to Periodicals* cover feisty metropolitan or statewide weeklies such as New York City's *Village Voice*, Washington D.C.'s *City Paper*, or North Carolina's *The Independent*, as well as defunct counterculture newspapers which, in their heyday, uncovered vast quantities of scandalous (and still relevant) material about people in high places. If you are investigating a crooked politician, landlord, or businessperson, the alternative press is often the best place to start (the *Village Voice*'s investigative journalism, for instance, is head and shoulders above that of the heavily self-censored *New York Times*).

## 6.14 Ethnic Weeklies

Virtually every ethnic group in America has its own weekly or weeklies, which give detailed coverage of the achievements of its members. Jews and African Americans have scores of local newspapers in all parts of the country. Smaller ethnic groups may have one or two papers giving nationwide news; for instance, the New York-based *India Abroad*, which has regional editions in Chicago, Toronto, and Los Angeles and which skewers any member of the Indian-American community who gets convicted of a crime. (*India Abroad* happens to have a very well-organized clippings morgue.)

Gale Research Inc. publishes the *Index of the Black Press*, which includes ten newspapers, mostly weeklies. In addition, abstracts of articles from Black newspapers can be accessed via UMI's Newspaper Abstracts database (1984 to 1988) and UMI's Newspaper and Periodical Abstracts database (1989 to present), both available on DIALOG.

## 6.15 Professional and Trade Publications

Articles that feature or mention your subject may have appeared in a professional or trade publication. If it's an interview in a publication read only by subject's colleagues, he or she may have spoken far more freely than if being questioned by a reporter from the major media. To find the most likely publications, look under the appropriate topic headings in *Associations' Publications in Print, Standard Periodical Directory*, and *Directory of Newsletters*. Also check *Ulrich's International Periodicals Directory* (subject may have talked most frankly of all to a foreign trade publication).

To find additional trade and professional publications, look in the *Encyclopedia of Associations* and the *National Trade and Professional Associations of the U.S.* If an association is listed, it probably has, at the least, a newsletter for its members. Simply call the association's research director and ask for the names of its local, regional, and national publications and where library backfiles might be located.

The most important trade and professional newspapers and periodicals are indexed in such volumes as the *Business Periodicals Index* or can be accessed online via major databases such as NEXIS. Others must be searched issue by issue, unless they include an annual index at the end of each year. Fortunately, many have a short section in each issue devoted to news about members. If your subject is prominent in his or her trade or profession, it may repay you to spend an hour or so going through back issues.

If you don't find any direct information about your subject in a professional or trade periodical, you will at least find the names of officers of the association, prominent members, and others who might have valuable leads. You may also find some potentially significant details about subject's employer or one of subject's closest colleagues.

## 6.16 House Organs

The in-house publications of companies for which subject has worked may include noncontroversial background information on his or her career and personal life. Such publications tend to be only four to twelve pages long and give most of the personal news about employees in a single section; it is thus easy to search through several years' worth in a short time. House organ news items may tell you if and when your subject was married, births and deaths in his or her family, and job promotions or professional honors he or she has received. Equally important, house organs will give you the names of many of subject's past and present employees, superiors, or co-workers.

Your gateway to the world of house organs is the *Internal Publications Directory* (formerly the *Gebbie House Magazine Directory*), which is part of the *Working Press of the Nation* series. This book describes 3,800 house organs (by no means a comprehensive list). Current issues of a house organ can sometimes be obtained from the company's public relations department. For back issues in libraries, check the *Union List of Serials* and *New Serial Titles*.

# 6.17 Alumni Newspapers and Magazines

If you know what college your subject attended, contact the editor of the alumni paper at the alumni office on campus. Many of these papers maintain clippings files of back-issue articles. As boosters of their school they are happy to furnish copies of any article that shows what high achievers their alumni are. Even if they don't have clippings files, the editor may recall an article on your subject. An alternative approach is to contact the university archives, which may have clippings files more comprehensive than those in the alumni office.

In searching the bound volumes of an alumni magazine, note that brief biographical notes and obituaries are often organized under class headings. If subject is a member of the Class of 1956, you can quickly search through the listings for that class in each issue.

# 6.18 Trade Union Publications

In the United States and Canada, there are hundreds of labor union newspapers, magazines, and newsletters published either by international unions or by state or local AFL-CIO councils, or by individual unions on the district or local level. In some unions, each local will have its own publication. District councils (the locals of a given union within, say, a given metropolitan area) may also have a publication. The backfiles may have valuable information on your subject if he or she is a union officer, an active rank and filer, or a management figure who has clashed with the union.

Don't neglect the dissident newspapers put out by rank and file groups at odds with the union bureaucracy. For instance, *Convoy Dispatch*, the Teamsters for a Democratic Union monthly, is an excellent source on the misdeeds of Teamster officials on every level.

Collections of trade union newspapers can be found in university libraries, especially if the university has an industrial relations department. To find the nearest library with back copies of a particular publication, see the *Union List of Serials* and *New Serial Titles*.

## 6.19 Sports, Hobby, and Other Specialty Publications

Is subject a collector of ancient coins? A breeder of prize-winning dogs? A rock climber? An ardent participant in bridge tournaments? Look in *Associations' Publications in Print* as well as *Standard Periodical Directory* for obscure periodicals relevant to subject's field of interest. Journalist Steve Weinberg found interesting material for a biography of billionaire Armand Hammer in publications as obscure as *Arabian Horse World*.

## 6.20 Publications of Fraternal, Civic, and Charitable Organizations

If a biographical dictionary lists subject as holding membership (or a volunteer or paid position) in any fraternal, civic, or charitable organization, look in *Associations' Publications in Print* for the name and address of that group's newsletter or bulletin. If no publication is listed, inquire directly from the association's national or local headquarters. Note that many associations (e.g., the Boy Scouts) have state or regional as well as national publications; always check out the one for subject's locality.

## 6.21 Religious Publications

If subject is in the clergy or active in lay affairs, his or her activities are almost certain to have been covered in denominational publications on some level. Look especially in state or regional publications of the given denomination.

Weekly church bulletins handed out at the Sunday services often include news about members of the congregation, e.g., births, marriages, participation in a mission work team, election to the church governing board, or appointment as a Sunday school teacher. If one of subject's children is being baptised or confirmed on a particular Sunday, that too will be in the bulletin. For many blue-collar families, these publications may be just about the only place they are mentioned in print. Church bulletins are usually on file in the church office, often going back decades.

For news of the clergy and prominent lay persons, you might check the online church news services. These include Lutheran News Service, RNS Daily News Report, United Methodist Information, ChurchNews International, and Catholic News Service, all available from CompuServe and/or NewsNet.

## 6.22 Genealogical Periodicals and Indexes

If there is a periodical devoted to persons with subject's surname or subject's mother's maiden name, it may contain biographical information about subject or some of his or her relatives. *The Directory of Family One-Name Periodicals* lists 1,600 of these publications. Many other genealogical periodicals are indexed (by individual as well as family name) in the *Periodicals Source Index (PERSI)*, which also covers local history periodicals; always check the *PERSI* supplements as well as the base set.

## 6.23 The Reporter—His or Her Sources and Files

The main object of searching through newspapers and periodicals is not simply to compile more and more clippings; rather, it is to find live sources: the people behind the news stories who know the things that didn't get printed.

Clippings will lead you to basically four types of people: the reporter who researched and wrote the article; the people mentioned in the article as participants in the reported events; the people whom the reporter quotes as sources (whether participants, eyewitnesses, or experts with background information); and the reporter's unnamed informants.

If the reporter is an expert on your subject (for instance, the longtime reporter on the labor beat who has written several articles about the Teamster official you are investigating), get whatever background information and advice he or she is willing to provide and, if possible, gain access to his or her private clippings files and a referral to his or her chief sources. (Contrary to the TV depiction of reporters, most are not jealous of their files and sources unless a major scoop is involved. In approaching them, remember that it's in their interest to cooperate with you if your research can fill in gaps in their own work.)

Sometimes, the reporter may not know very much—the article was only one of many hurried pieces written on a tight deadline. In such cases, you may want to see whatever documentation the reporter has retained, but your main objective will be to get the telephone numbers of the sources who provided most of the information.

## 6.24 The Amateur Muckraker

This is the freelance writer or citizen-researcher devoted to gathering all the scandalous clippings on everyone in town (or at least on particular groups or individuals who have incurred his or her wrath). At their best, muckrakers can be an almost miraculous source of information on evildoers, and you should urge local reporters to put you in contact with them. For a full discussion, see 13.1.

## 6.25 Library Clippings Collections

Hundreds of libraries around the United States have newspaper or periodicals clippings files donated to them by private researchers or compiled by library staffers. A good example of the treasures you might find is the vast collection of news clippings in the North Carolina Collection at the University of North Carolina's Wilson Library in Chapel Hill. The library staff began collecting these clippings from newspapers across the state in the 1920s. Photocopies of all clippings through 1975 have been compiled into 364 volumes, which are divided into biography and subject collections. The biography volumes, organized alphabetically by name, contain clippings on tens of thousands of North Carolinians. They include profile articles, interviews, and obituaries as well as news articles centered on the individual's activities. You can access further information on the individual by looking in relevant subject files. The library staff is currently preparing a second set of volumes to cover the years 1976-1989.

For information on how to find similar collections at other libraries across the country, see 11.2. For information on private and organizational clippings files, see 13.4.

# 7 ∙

## Collecting the Basic "Identifiers"

To conduct a thorough background check, three items are most important: first, subject's full name correctly spelled, including full middle name and correct generational designation (e.g., "Jr.," "III"); second, subject's date of birth; third, subject's social security number (SSN). Without the first, it is sometimes difficult to follow the paper trail even on the simplest level (especially if the name is a common one). Without the second and third, it is difficult to find public records filed according to these identifiers.

## 7.1 What's in a Name?

Without any deceptive intent, many individuals leave a confusing paper trail because of marital name changes or informal name variations. When you first see your subject's name in a newspaper article or phone directory or on a mailbox, you should not assume this is the full name under which most records regarding subject's past are filed. Indeed, there may be no one form of his or her name that covers most of the available documentation.

### Middle Names, Nicknames, and Aliases

Your subject may be commonly known by his or her middle name, a nickname, or a shortened form of his or her first name (e.g., "Dell" for "Delmore"), and may give any of these as the first name in a telephone listing or when introducing himself or herself. John Quincey Public may receive utility bills as Quincey Public, receive Mastercard bills as John Q. Public, sign his name on checks as J. Quincey Public, and be listed in the phone book (and also be known to most acquaintances) as "Quince" Public. In searching through phone or crisscross directories and county courthouse indexes, this might only be a minor annoyance since "Public" is such an uncommon surname. But if your subject's last name is Smith, you will to have to get things clear. This is all the more necessary if subject has more than one middle name and varies their use according to whim, e.g., John Gerald Wellington Marshall Smith, who is always one step ahead of the bill collectors!

In your earliest interviews with persons who know subject, find out whether he or she is usually called by first or middle name, what his or her past and present nicknames are, and whether he or she has ever used any pseudonyms or aliases. Also note carefully any evidence of name variations in newspaper clippings about subject.

### Generational Designations

Confusion may arise when father and son have the same first and middle as well as last name. This is supposed to be cleared up by the use of "Jr.," "III," and "IV." But your subject and those with whom he shares the name (grandfather, father, son, or grandson) may not use the generational designations consistently. For instance, if the father and son live in different cities, the son may not bother to include "Jr." or "III" in his telephone listing. Or he may drop the "III" because it sounds pompous. Someone searching for the son will see "Jr." after the father's name in the phone book, and mistake father for son. Confusion may also result when father and son live in the same household and the telephone listing for one or the other lacks the proper designation.

### Maiden Names, Married Names, and Other Variations

If a married woman uses her husband's surname, you will still need her previous name or names. Records from before her marriage will be listed by her maiden name and/or previous married name, and she may still use her maiden or previous married name in her professional career. (Note that her maiden name is not necessarily her birth name—she may have taken it from her stepfather when her mother remarried.) If subject is her husband, he may be using her maiden name (or the name of one of her parents) as the "straw name" to conceal his ownership of a real estate parcel.

In this era of frequent divorce and remarriage, joint custody of children, two-career households, and legally recognized gay and lesbian partnerships, name variations can become extremely complicated.

- Wives frequently use hyphenated surnames, and the practice has been adopted by some husbands. The wife may put her name first, the husband may put his first, she may put his first, he may put hers first, both may put his first, both may put hers first. In addition, they may vary the usage according to the situation or their mood. In filling out a job application, for instance, the husband may drop the hyphenated name to avoid appearing flakey.

- The wife may retain her maiden name for all purposes, social and business, and may be listed in the phone book ONLY under that name. Or she may shift from maiden name to married name (or vice versa) for certain purposes (e.g., to establish an alternate credit history if she has a bad credit rating under the other name).

- When a teenage child is living with a mother who has remarried, the mother may use her new husband's surname while the teenager continues to use (and has a telephone listing under) the father's surname. Or, the teenager may use the father's name in some situations and the wife's new husband's name in other situations (for other variations on this, see 4.11).

- Parents may choose a surname in the maternal line (or a hyphenated surname with either husband's or wife's name first) for a child at birth—this is an increasingly common practice since the early 1980s.

- Parents may chose a surname at random for their child. According to the *Wall Street Journal*, February 11, 1987: "Parents are making use of little-known laws that allow them to bestow on the children the surname of their choice. Short of a curse word or a series of numerals, the choice in most states is unlimited."

- When an unmarried couple is living together, the woman may use the man's surname or a hyphenation of the two names in certain social situations or in signing an apartment lease (hence ensuring that the "married" name will appear in the building lobby's directory). But she may continue to use her maiden name or the name of the husband from whom she is separated or divorced in other situations. In the case of one couple I tracked, this was complicated by the fact that both were using "political" surnames at meetings of and in their writings for an extremist sect. Over a ten-year period, the woman used, interchangeably, her lover's political surname, her lover's real surname, her previous married surname, her previous political surname, and her maiden name.

### Common-Law and Statutory Name Changes

If you are tracing subject's name backwards in old telephone books and the trail runs out, it could be that subject has changed his or her name. Emigrants of earlier generations often "Americanized" their names. Black Americans often abandon their "slave names" for Arabic, West African, or Swahili names. Persons who have suffered public disgrace may change their names to facilitate building a new life. Actors may adopt names that will enhance their box office appeal. And some people change their names on a whim.

The laws regarding names changes vary from state to state. In New York, the right to change one's name for honest purposes is a common-law right, which can be exercised by simply beginning to use the new name in all transactions and having it recognized by friends and associates. This right, which includes that of the mother of an illegitimate child to adopt the name of the putative father for herself and the child, does not require court permission. However, the law provides for statutory name change (by order of a court) as an affirmance of the common-law right. This process is distinct from name changes pursuant to marriage, adoption, divorce or annulment.

Court approved name changes are typically recorded at the county courthouse in a ledger book that gives both old and new names. The number of name changes per year is usually quite small.

Whether a person uses the court method or the common-law method, he or she will have to notify various ID issuing agencies, his or her bank, etc. When a person notifies the Social Security Administration of a name change, that change is recorded but the person is NOT issued a new SSN. The SSN thus becomes a convenient means by which an investigator can establish (via credit-reporting agency records) the link between subject's old and new names.

Note that some people who adopt a new name (especially those who do so via the common-law use method) may only selectively notify public and private record-keeping agencies, thus continuing to use the old name in various transactions. Even if a person does notify the most important agencies, there will be a time lag before the new name fully replaces the old.

If you can't find any record of a name change, the person with the suspiciously short paper trail may be using a false identity based on fraudulently obtained ID (see 7.5).

### Finding Subject's Previous and Alternate Names

College alumni directories and class anniversary directories give both the maiden and married names of former women students. Society guides (such as *The Social Register*) also provide both names, as do family name books, genealogy charts, and the biographical dictionary entries for women. Biographical dictionary entries for men often give the maiden names of their wives.

Alumni directories, class anniversary directories, high school and college yearbooks, and newspaper clippings may provide subject's former or present nicknames.

The judgment docket at your county courthouse sometimes lists a/k/a's beside a person's name—these may include personal aliases as well as business names and property ownership "straw" names.

Criminal court records and newspaper crime articles may give the aliases of criminals.

### Making Sure You Have The Right Person

In the early 1980s, Edward H. Heller of Brooklyn practiced law at 230 Park Avenue in Manhattan. Meanwhile, another Edward H. Heller, also of Brooklyn, practiced law at 250 Park Avenue. When the former was convicted of grand larceny and disbarred, the latter wrote a letter to the *New York Law Journal* to clear up the confusion, signing his name "Edward Harris Heller."

The innocent Mr. Heller's problem was no different in essence from that of the many Americans who are mistakenly arrested each year—or denied credit—because of name confusion in databases rife with small errors that are really big errors.

Never assume you have the right person unless (1) the middle name (not just the middle initial) fits; (2) the address (including street number) fits; (3) the date of birth and SSN fit (the latter are especially important in avoiding mistakes in generational designation, as when the John Smith of 221 Elm Street, arrested for indecent exposure, turns out to be John Smith Jr., the emotionally disturbed son who lives in John Smith Sr.'s basement).

# 7.2 Obtaining the Birth Date

### *"Real" Versus Operative Birth Date*

In following a person's paper trail, you can become confused if you don't distinguish between "real" birth date and operative birth date. The "real" date is that found on the birth certificate. (I have put "real" in quotation marks because, prior to the sexual revolution of the 1960s, dates on birth certificates were often fudged to disguise the fact that a child was born less than nine months after the parents' wedding.) The operative birth date is that which is used in adult life as an identifier. The two are frequently different because there are so many reasons to deliberately misreport one's age. The initial misreporting may occur early on, as in the case of teenagers who added a year to their age in order the join the Marines during World War Two. More often, people misreport their age later in life to avoid age discrimination when applying for a new job, to retire or qualify for benefits early, to delay retirement, or simply for vanity's sake. Often, the "new" birth date may become operative in one set of records but not in another (say, employment applications for successive jobs, but not applications for new bank accounts). If the "new" date is used on a credit application, it may be part of an attempt (also involving a shift from married to maiden name, the listing of a new address and phone number, etc.) to set up an alternate credit history.

The operative date of birth is best defined as that which is recognized by any particular records system as an identifier for subject's records. Below, unless we are speaking of the birth certificate, date of birth will be used in this sense.

## Getting The "Real" Birth Date

In the county courthouse of the county where subject was born there may be a birth index that you can look through to find the date of birth. This birth index may also be on microform in the genealogy division of the local public library or the Mormons' local Family History Center. The New York Public Library's genealogy division, for instance, has the city health department's annual birth indexes from 1917 to 1982.

If you know the state but not the county in which a person was born, contact the state's bureau of vital statistics (usually part of the state health department). In many states, you can get the birth records searched and obtain a copy or abstract of the birth certificate. For more information on how to deal with county and state vital statistics registrars, see 10.1.

A subject's "real" date of birth may also be found in church baptismal and christening registers, which are usually open to the public. Many church registers from all parts of the country have been collected on microform by the Mormons.

## Credit-Reporting Agency Databases

The operative birth date is header material on credit files and as such should be readily available from credit-reporting agency databases via an information broker.

## Biographical Dictionaries and Professional Directories

Although some of these works provide only the year of birth, others (which include collectively the names of millions of Americans) provide the full date. For instance, *American Men and Women of Science* generally includes the date of birth in its 125,000-plus entries. *The Martindale-Hubbell Law Directory* gives only the year of birth in its roster listings, but includes the date of birth in biographical sketches of about 40,000 attorneys.

Note that with the year of birth alone you can still search many records systems (for instance, some military personnel records)—it's just a bit more difficult. Depending on the system being searched, secondary identifiers such as place of birth and mother's maiden name may compensate for the lack of a more precise date.

## Miscellaneous Public Records

A subject's date of birth will be on his or her marriage license. It will usually be on driver's license abstracts and in voter registration records.

## Military Rosters

Date of birth is included in the annual registers of Army, Air Force, Marine, Coast Guard, and Army National Guard officers. The Navy register gives only the year of birth.

### Applications For Jobs, Loans, etc.

A date of birth, accurate or not, will be on every application subject has ever filled out for a job, loan, rental of a house or apartment, or admittance to a college. A landlord or former employer may be willing to dig out such an application from his or her files.

### Date Of Birth: No Great Secret

Some investigators will call a subject, say they want to include him or her in an occupational directory, and get various basic items (including birth date) in a few moments. This type of deception is not recommended—the story simply illustrates the point that, because date of birth is so commonly asked for, most people are not secretive about it.

# 7.3 Finding the Social Security Account Number

As header material on his or her credit file, subject's SSN is readily available from credit-reporting agency databases via an information broker. The broker may need to check multiple database systems to find it—the price can range from about $15 to $35 depending on how many databases are searched. In many cases, however, you won't need to pay a broker— you can easily obtain the SSN while checking various public or private records that you would check anyway. What follows are some of the places the SSN might be found.

### Driver's License Abstracts

In some states, the SSN is always the same as the driver's license number or is listed on the driver's license in addition to the driver's license number. In other states, the use of the SSN is optional (for instance, in Massachusetts, where about 15 percent of drivers request a non-SSN number) or it is not used at all.

If your subject lives in, say, New York State, where licenses do not include the SSN, you might check the records of lapsed licenses or license transfers in the state in which subject previously lived. (For more on driver's licenses, see 10.2.)

### Voter Registration Records

In some localities the SSN is included on your voter registration card as a safeguard against vote fraud. Voter registration records are kept at the local board of elections and are almost always open to the public. The use of the SSN for this purpose has declined in recent years, so look in older records if available.

### Court Records

The SSN may be found in civil court records; for instance, when subject's tax returns or other financial documents are attached as exhibits to a motion or affidavit. If subject has ever been convicted of a crime, you may find his or her SSN in the criminal court file. If a person has ever filed for personal bankruptcy, the case file may include his or her SSN.

### Disclosure Statements and Other Filings
The SSN may be included in financial disclosure statements required from elected and appointed public officials, high-ranking civil servants, and candidates for public office.

In addition, you may find the SSNs of officers of nonprofit corporations on the 990 forms filed by their organizations with the IRS.

### Applications For Jobs, Housing, etc.
The SSN, like the date of birth, is required information when a person opens a bank account, rents an apartment, or applies for a job, loan, mortgage, or credit card. If you are investigating an obnoxious local cult, you may find that local landlords and businesspersons will give you access to cult members' apartment rental or job applications.

### Bank Statements and Other Personal Financial Records
The SSN is sometimes found on subject's bank statements, health insurance bills, statements from money market funds, etc. Canceled checks may have the SSN (or driver's license number including the SSN) scrawled on the back if subject had to show ID to cash the check. The SSN, since it is used as a person's taxpayer identification number, will be on all correspondence he or she receives from the IRS. Subject may routinely throw such things out with the garbage. The use of garbology as a research technique is described in 11.4.

### Military Records
Present, former, and retired members of the Armed Forces, Reserves, and National Guard are identified by their SSN. Although the Privacy Act of 1974 prevents the government from giving out the SSN when you request someone's service record, you may find the SSN on discharge papers filed at the county courthouse.

Until the mid-1970s the SSN was listed beside each officer's name in the register of officers for each service. The Army National Guard register included SSNs as late as 1980. Back issues of the service registers may be found at your nearest federal depository library.

### Multiple "SSN" Holders and Other Special Problems
Under Social Security regulations any SSN holder may change his or her SSN by applying to the Social Security Administration and showing good reason. Some people do it because of suspicion that their SSN and name are being used by someone else for fraudulent purposes. Others do it to put distance between themselves and negative credit information held in databases. Militant trade unionists in past decades did it to avoid blacklist-

ing. But after a person obtains the new SSN, the old one does not disappear suddenly from non-Social Security Administration records. For instance, if the old SSN was already established as the identifier on subject's health insurance policy, it could continue as such indefinitely.

The problem of tracking people by their SSN is compounded by the many deceptive practices involving this so-called universal identifier. Some people will switch the digits on their SSN or make up a fictitious one if they are opening a new bank account that they don't want bill collectors or the IRS to know about. Others will list SSNs taken from stolen or forged Social Security cards (illegal aliens frequently do this). Con artists have been known to expropriate a stranger's identity and SSN (as by learning the SSN and other identifying information through a telephone ruse, then applying for a new card in the victim's name), so that John Doe who has an impeccable credit rating in Connecticut becomes John Doe the deadbeat in Oregon. Participants in the so-called underground economy may utilize a false identity (and an SSN issued in that name) for certain banking transactions while continuing to use their real identity and SSN on other occasions. The above are only a few of the possible frauds and deceptions that make it difficult to track people by their SSN alone.

## 7.4 Making Use of the "SSN" Code

Have you been unable to find out where subject grew up? Do you suspect that subject has given false information about his or her past? The SSN may be the key to unraveling the mystery.

The nine-digit SSN is divided into an area number (first three digits), a group number (middle two digits), and a serial number (last four digits). Both the area number and the group number are useful for our purposes.

### Area Number
The Social Security Administration assigns the area number based on the mailing address given by the applicant. The area number will fall within a consecutive series of numbers that identify the state in which that mailing address is located. With a single minor exception (see below) no two states share the same range of area numbers.

The majority of today's adult native-born Americans obtained their SSN in high school or shortly thereafter. Subject's SSN will thus usually indicate the state in which he or she grew up or at least a state in which he or she once resided. If the last name is an uncommon one, you can check the regional white pages database (see 4.4) or Directory Assistance for each area code in the state to see if subject still has relatives living there. Or you can check the files of lapsed or transferred driver's licenses in that state.

(Note that as of 1987, federal law requires parents to obtain SSNs for children five years old or older who are listed as dependents on income tax returns. Beginning in 1992, this requirement will apply to children one year old or older. The inferences one would draw from area numbers should be adjusted accordingly.)

Even if you don't want to bother with tracing your subject's paper trail to another state, the SSN will at least give you an indication of whether or not subject is being truthful about his or her past. If subject claims to have grown up and attended college in Hawaii, but the SSN is coded for North Dakota, you have definite grounds for suspicion.

The following is a list of area numbers and the states to which they correspond:

| | |
|---|---|
| 001-003 New Hampshire | 440-448 Oklahoma |
| 004-007 Maine | 449-467 Texas |
| 008-009 Vermont | 468-477 Minnesota |
| 010-034 Massachusetts | 478-485 Iowa |
| 035-039 Rhode Island | 486-500 Missouri |
| 040-049 Connecticut | 501-502 North Dakota |
| 050-134 New York | 503-504 South Dakota |
| 135-158 New Jersey | 505-508 Nebraska |
| 159-211 Pennsylvania | 509-515 Kansas |
| 212-220 Maryland | 516-517 Montana |
| 221-222 Delaware | 518-519 Idaho |
| 223-231 Virginia | 520 Wyoming |
| 232-236 West Virginia | 521-524 Colorado |
| 237-246 North Carolina | 525, 585 New Mexico |
| 247-251 South Carolina | 526-527 Arizona |
| 252-260 Georgia | 528-529 Utah |
| 261-267 Florida | 530 Nevada |
| 268-302 Ohio | 531-539 Washington |
| 303-317 Indiana | 540-544 Oregon |
| 318-361 Illinois | 545-573 California |
| 362-386 Michigan | 574 Alaska |
| 387-399 Wisconsin | 575-576 Hawaii |
| 400-407 Kentucky | 577-579 District of Columbia |
| 408-415 Tennessee | 580 Virgin Islands |
| 416-424 Alabama | 580-584 Puerto Rico |
| 425-428 Mississippi | 586 Guam, American Samoa, and all |
| 429-432 Arkansas | other Pacific Territories |
| 433-439 Louisiana | |

Three special cases fall outside the above categories. First, the area numbers 700 through 728 were assigned, prior to 1963, to railroad workers covered under the Railroad Retirement Act, irrespective of the state in which the worker applied. Second, the California numbers 568-30 through 568-58 were issued to Vietnamese and other refugees between 1975 and

1979. Third, the area number 232 with middle digits 30 was reassigned from West Virginia to North Carolina.

### Group Number

The fourth and fifth digits—the group number—can sometimes give you an indication of when the SSN was issued. This will help you pin down the time during which your subject resided in the state identified by the area number. It also may suggest that the individual is using false ID or that something odd has been going on in his or her life: If subject is a man in his mid–fifties, why should he be using an SSN issued no earlier than 1972? If this is his first number, how was he able to get and keep jobs in his earlier life without having a number? If he obtained a new number to replace the old one, what was his motive?

The assigning of group numbers is a complicated matter, and cannot be explained properly in a few paragraphs. The National Employment Screening Services' "Social Security Number Guide" (see bibliography) explains how to figure out if a job applicant has listed on his application an SSN that includes a group number not yet issued by the SSA. The chapter on SSNs in the SSA's *Program Operations Manual System* (*POMS*), available at your nearest SSA office, will tell you such esoterica as how to spot an SSN that was issued prior to 1972 (the year in which SSN issuance by field offices was replaced by a system of central issuance from headquarters in Baltimore).

## 7.5 How to Detect a False Identity

If records of subject's past go back only a few months or years, it may be that he or she is "paper tripping," i.e., using a false identity based on the birth certificate of a dead person. Paper trippers will usually pick the identity of someone who died in infancy or early childhood, and whose date of birth and basic physical characteristics, as recorded on the birth certificate, are similar to those of the imposter. The imposter obtains a copy of the birth certificate and uses it to obtain a Social Security card, a driver's license, and other ID. This is possible in many localities because applicants for birth certificates do not have to show proof of identity and also because there is no cross-register of birth and death certificates.

Careful paper trippers will select the identity of a dead person who was born in one state and died in another; this ensures them against being unmasked as a result of any future statewide correlation of birth and death records. To guard against accidental discovery (as by an encounter with a sibling of the dead person), they may search through old newspapers to find reports of accidents in which entire families perished (e.g., an auto crash or a household gas leak or fire) and then adopt the identity of one of the victims.

One indication that a person may be an imposter is if his or her SSN number does not fit plausibly with his or her reported age (e.g., the man in his fifties who has an SSN issued after 1972). As noted above, many people obtain a legitimate new SSN to replace an old one; but if so, the credit-reporting agencies will have a record of both—and there will be other evidence of a paper trail for the person in question dating from before the new SSN's issuance.

Another tip-off is if subject received a driver's license for the first time at an age older than usual. One would then look in the white pages and/or crisscross directory: Did subject's name first appear in the directory only about the time he or she received the delayed driver's license?

In tracking an imposter, always check the central death index in the state in which he or she claims to have been born. Although (as noted above) many paper trippers prefer to use the identity of a child who was born in one state and died in another, this is not always easy to do (the majority of families who move from one town to another stay within the same state). Your subject may have been careless about this.

Next, try to find the parents and/or other relatives of the presumed dead child. Once you have the child's date and place of birth (which may be on the imposter's driver's license abstract or in credit-reporting agency header material), simply check the birth register or obtain your own copy of the birth certificate to find out the parents' names (see 10.1). Track them down (or track down surviving siblings or other relatives) using the techniques described in 4.11. In most cases these people will be indignant about the desecration of the child's memory, and will cooperate in exposing the imposter.

# 8.

# Credit and Financial Information

## 8.1 Credit-Reporting Agencies and Other Database Sources

The amount of information collected by credit-reporting agencies is staggering. TRW's Updated Credit Profile database, alone, has information on 133 million people, including data on their credit card payments, lines of credit, and secured loans. It also has public record information on tax liens, judgments, and personal bankruptcies. Equifax, another major national credit bureau, has tens of millions of consumer files that include gossip from a subject's neighbors (privacy advocates claim that up to one third of this information is inaccurate) as well as arrest and conviction data. Regional or local credit bureaus (of which there were about 1,300 as of 1988) often have more detailed files than either TRW or Equifax on individuals and small businesses in their locality. Nationwide, the credit-reporting industry has compiled over 400 million consumer credit files.

The Fair Credit Reporting Act sharply restricts the dissemination of credit information about individual consumers. To obtain such information, a client of a credit-reporting agency must have a substantial business need. Banks, department stores, insurance companies, credit card issuers, employers, and landlords fall into the category of clients with legitimate needs.

In spite of the law, private investigators and skip tracers routinely obtain credit information on consumers. This is often done through one of subject's creditors, since under the law any company can obtain a credit report on one of its debtors without obtaining the latter's permission (once the company has the report it can show it to anyone it chooses). Private investigators also make direct information-trading arrangements with people in the credit-reporting industry. This is easy to do, because many P.I.s previously worked for credit-reporting agencies, or, as police officers or insurance claims investigators, developed contacts at these agencies. In addition, there are allegedly organizations that sponsor social get-togethers of credit bureau employees and P.I.s to facilitate this illegal swapping.

Newspapers generally will have an account with TRW or some other credit-reporting firm. Although the newspaper will use this service mostly for business purposes, such as checking out job applicants, reporters have been known to use it to gather information on their investigative targets.

Be aware that under the Fair Crediting Reporting Act anyone who knowingly obtains information on a consumer from a credit bureau under false pretenses can be fined up to $5,000 and imprisoned for up to one year. Several states also provide stiff criminal penalties. In addition, anyone caught obtaining a consumer credit report illegally can be sued by the consumer.

## 8.2 Alternatives to the Credit-Reporting Agency

There are three basic ways to compensate for the restrictions on consumer credit reporting while staying completely within the law and adhering to journalistic ethics. First, the portions of the credit-reporting agency file on an individual obtained from public records (e.g., UCC filings and judgment docket listings) are available directly from the state or from commercial firms that are not part of the credit-reporting industry. You can obtain this information yourself at the county clerk's office, or you can order certain records from the county or the state by mail or over the phone, or for speedy delivery you can call a documents search service.

Second, the Fair Credit Reporting Act does not restrict the dissemination of business credit information. Thus, if you are backgrounding a businessperson, you can run a database check on his or her known business entities. Dun & Bradstreet's credit reports on over nine million business locations are available online to direct subscribers. TRW business credit reports and some D & B reports are available throughout CompuServe or NewsNet (the latter offers trade payment histories on over thirteen million business locations). This business credit information may give you a better picture of subject's overall financial status than any individual credit check would provide: An individual may have a scrupulous record of paying his or her household bills on time—yet his or her business may be on the brink of bankruptcy. For more information on backgrounding a business, see Chapter Twelve.

The third way to compensate for consumer credit-reporting restrictions is simply to do some leg work—dig into real-estate records at the register of deeds office, search the plaintiff/defendant index and case files at your local courthouse, interview former associates of your subject, etc. In doing this, you may miss some things that are in the credit bureaus' files. But you will collect much information—especially from little-known public records the credit bureaus almost never consult—that may give you an excellent picture of subject's finances.

## 8.3 Uniform Commercial Code Filings

Whenever a person borrows money from a bank, finance company, savings and loan association, etc., and offers personal (non-real) property as collateral, the lender fills out a Uniform Commercial Code (UCC) financing statement and sends it to the state's Department of State. A copy may also be filed with the county clerk's office or the register of deeds office in the county where the transaction occurred.

Any quick search at the county level should be followed by a search of the statewide files. In some states you can get the information over the phone either for free or for fees ranging up to $25 (these states are: Colorado, Florida, Idaho, Iowa, Kansas, Michigan, Mississippi, Missouri, Montana, Nebraska, Ohio, South Dakota, Texas, Utah, West Virginia, Wisconsin, and Wyoming). In most other states (and in the above states as well, if you want a certified copy), you must request a search in writing using a UCC-11 form (Request for Search) and enclosing the required fee. Within a few days you will receive an abstract of all UCC financing statements statewide in which your subject is listed as a debtor. (If you need such searches often, you can purchase UCC-11 forms in bulk from Julius Blumberg, Inc. in New York City; the same form can be used for most states.) Note that in several states a search can only be ordered through a private title search company. (See *Lesko's Info-Power* for specific requirements in each state.)

Several states offer online access to UCC files. These include California, Colorado, Florida, Mississippi, Montana, South Dakota, Texas, Utah, and Washington. Through information brokers you can get easy access to these databases as well as to commercial databases compiled directly from the state UCC files in other states. If you go through LEXIS, its Lien Library provides UCC filings for California, Illinois, Maryland, Pennsylvania, and Texas; its LEXDOC service can obtain certified or uncertified copies from anywhere in the United States.

At the minimum you should check the UCC files for subject's state of residence and for any state in which subject conducts business or has a vacation home. If subject resides in a multistate metropolitan area, check each state (e.g., New Jersey and Connecticut should be checked for any New York City resident).

If you are checking the county-level files yourself, note that if a debt has been satisfied, the UCC statement will be discarded after a fixed period. If the debt is not satisfied, the statement will remain in the file indefinitely. The statement will not tell you the amount of the loan. It *will* tell you the names and addresses of debtor and creditor, the date on which the financing statement was filed, and whether or not the obligation has been satisfied (i.e., paid in full); it will also describe the asset offered as collateral.

The UCC filings should be regarded as a guide to further investigation. For instance, if the collateral offered by the borrower is an airplane or yacht, a number of relevant public records can be examined. If the collateral is a painting or statue, this may lead you into an examination of subject's relationship to the world of art dealers and museums.

You may find information about subject's real estate holdings in the UCC files; for instance, if fixtures in the building have been put up as collateral for a loan.

If the UCC filings reveal that subject obtained a recent loan from a particular bank, this may be the same bank at which subject has his main checking and savings accounts (see 8.18). If the loan comes from a sleazy finance company, this suggests that subject has a mediocre credit rating.

## 8.4 Judgment Books

Money judgments obtained against your subject in the local courts are usually on file at the county clerk's office. Recent judgments may be listed in a computer index; older judgments, in annual ledger books (with separate books for judgments against individuals and corporations). The ledger books are permanent records—you can access them going back as far as you like.

In New York City, each judgment book entry tells the amount of the judgment, the names and addresses of both creditor and debtor (and often of creditor's attorney), the date of filing of the judgment, the court in which the judgment was obtained, the docket number of the case, and the date of satisfaction (if any). If the judgment was also obtained against another individual or individuals, a partnership, or a corporation, this will be noted in the entry for your subject. These books also include judgments obtained by the city or state tax commissions; IRS judgments are filed separately in federal tax lien files (see 8.5 below).

The judgment books are a rich source of leads. For instance, the individuals listed along with subject as the targets of a judgment may be business partners you had not known about. A judgment obtained against subject by a hospital may be your first clue as to his or her serious medical condition.

If the State Attorney General is listed as the creditor, this may mean the state has obtained a judgment against subject because of nonpayment of a fine resulting from an enforcement action penalizing him or her for illegal activity.

Other judgments may provide the names of subject's former customers, clients, vendors, or spouses. In one recent investigation, I checked the Manhattan judgment books going back ten years and found the name and address of a woman who had obtained a judgment for several thousand dollars against subject (a Manhattan restauranteur) in a dispute over an item of jewelry she had lost in his restaurant. As it turned out, she and the restaurant owner had many friends and business associates in common; I was regaled with extremely interesting gossip.

The judgment books in the county clerk's office will cover actions in state district court (often called superior court) and in the county or municipal courts; the books may also record judgments from other localities. Federal court judgments will be recorded at the federal district court. Once you have the docket number, you can obtain the case file.

Note that a judgment may be obtained as part of a fraudulent conspiracy between creditor and debtor. Let's say that Mister X has borrowed large sums with the intent to declare bankruptcy and thus evade payment. He gets a crony to obtain a large judgment against him. When bankruptcy is declared, the crony has a secured prior lien and gets paid first out of money that otherwise would go to legitimate creditors after tax debts are satisfied. Later the crony returns the money to Mister X minus his or her own cut.

## 8.5 Federal Tax Lien Files

These may be located at the register of deeds office or county clerk's office. Filed alphabetically by debtor's name, they give the date of perfecting (e.g., the date the lien was obtained), the amount owed, and the date of satisfaction (if any). Corporate and individual tax lien records are generally kept in separate files. Like UCC statements, they are retained as long as the lien is unsatisfied. Once it is satisfied, the card will be removed from the file within a fixed period.

The size of tax liens will give you some clues as to subject's annual income. If Mr. X is known by you to have a modest-paying civil service job, but the IRS is after him for $50,000 in back taxes, this indicates a second source of income that you will want to track down. If Ms. Y has accumulated several liens over the last few years, it may indicate her catering business is barely keeping afloat. In addition, the addresses listed on the cards, as you trace the liens throught the years, may provide you with previously unknown former residential or business addresses of subject.

## 8.6 Wage Assignments (Wage Garnishments) Index

This index is usually located at the county clerk's office, with indexing by date and alphabetically by assignor (the person whose wages have been assigned). In New York City, the index gives the amount owed originally, the names of assignor and assignee, and a file number. This may lead you to a judgment obtained locally or to an out-of-town judgment. Depending on the locality, assignor's place of employment at the time of the wage assignment may be part of the public record. If not, you might obtain this information from the assignee.

## 8.7 Register of Deeds Office

Let's say your subject lives in a private house in a suburban community. Does he or she own it? If so, how much is the mortgage and who is the mortgagee? If subject doesn't own the house, who does? The key to answering such questions is the grantor/grantee indexes at the city or county Register of Deeds office. These indexes contain notices of mortgages, deeds of sale, property liens, etc. Each entry will give the names of grantor and grantee and the date and general nature of the transaction.

In New York City, you first get the block and lot number of the property from maps. Then you go to the most recent ledger book containing the records of that block and lot. Recorded with each transaction will be a file number, so that you can access the actual document from the microfiche files in an adjacent room. In this manner, you can examine every document pertaining to the history of the property. This will answer the questions posed above; it may also tell you who your subject's attorney is, where your subject does his or her banking (if it's the same bank that holds the mortgage), subject's spouse's name, what the signatures of subject and spouse look like, etc.

If your subject is an ordinary homeowner and not involved in real estate speculation, the information you dig up at the Register of Deeds office is not likely to be very exciting. Yet you may find clues to something important going on in subject's life. For instance, if subject took out a second mortage on his or her home in 1984, this may reflect business difficulties he or she was experiencing at the time, a major medical expense, the need to pay for a child's college education, etc.

## 8.8 Condominium and Cooperative Ownership Lists

TRW REDI Property Data publishes a number of local directories that contain information about condo and co-op owners. For instance, *The Record and Guide Quarterly* reports on condo sales in Manhattan, and includes both a buyer's and seller's index. TRW REDI's real estate directories covering New York City and other localities will include a listing of all condo owners alphabetically by name—you can thus see quickly if subject's apartment falls into this category (and if he or she got stuck with a bad investment when the real estate market crashed). As to cooperatives,

TRW REDI's directories for New York will tell if a building has gone co-op, but will not provide information on individual co-op owners. Since a large percentage of the occupants of most New York cooperative apartment houses are rental tenants left over from pre-cooperative days, or tenants renting from the co-op owner who bought the apartment as an income-producing investment, you shouldn't jump to conclusions one way or the other just because subject happens to live in a co-op building. However, if an occupant lived in the building before the co-oping occurred (which you can determine via the crisscross directory), he or she may have bought at an insider price. Depending on the present market value of the apartment, he or she may have more equity than one would think from looking at other personal worth and income indicators.

## 8.9 City or County Property Tax Records

These records are usually listed by date and by block and lot numbers. Once you have a list of the properties your subject owns, you should check the assessment rolls (land value and the total value for each lot), the tax abatement books (amounts of abatements for each block and lot), the tax rolls (assessed valuation, quarterly tax, total tax, abatements, arrears), and the tax registers (balance due, charges, payments).

If you are trying to calculate subject's worth, note that the assessed value of real property often varies widely from its market value.

Also, in calculating worth, don't forget subject's weekend or summer home, which will probably be located in another county.

## 8.10 Bankruptcy Court and Probate Court Records

If your subject has recently filed for relief under the bankruptcy laws, you're in luck. The records, kept in the bankruptcy division of the Federal District Court while the case is open (and thereafter at the nearest Federal Records Center), will provide a quite detailed picture of subject's income, liabilities, and assets, including personal property such as automobiles, stereos, and stamp collections.

Be sure to inspect bankruptcy records for corporations with which subject has been connected. These records may reveal his or her corporate salary at the time of bankruptcy, how much stock he or she owned in the corporation, etc. Chapter 11 bankruptcy proceedings are often the result of a well-thought-out scheme to defraud a corporation's vendors and other creditors. If your subject is a principal in the corporation, you should examine his or her role carefully.

The local probate court is also a vital source of financial information if subject has inherited any money or property. Wills offered for probate are kept on public file, as are other records pertaining to the protection and transfer of the estate. Important documents to look for include the estate appraisal and the reports filed by the administrator or executor regarding payments to the heirs. If the will is contested, much information will become part of the public record that otherwise would remain private. Probate court records are generally indexed by the name of the decedent; thus, to find all records that might shed important light on your subject's finances, you will need the names of any deceased grandparents, parents, in-laws, or other close relatives who were well-to-do, as well as the locations of the probate courts that handled these cases. (The Social Security Death Index is useful for this purpose.)

## 8.11 Court Records Of Civil Suits

Frequently, court records of civil suits will be the richest single source of information about an individual. See Chapter Nine for a description of the different types of court records and how to search them.

## 8.12 Federal and State Income Tax Returns and Related Information

If the IRS has settled a claim against subject for less than the amount originally owed, then IRS Form 7249-M will be filed at the regional IRS office for one year and thereafter in Washington. This form will contain rather detailed information about subject's salary and other income, assets, and liabilities.

If subject's dispute with the IRS has found its way into the U.S. Tax Court in Washington, the case file will contain a wealth of financial documentation, including income tax returns, affidavits, depositions, and court testimony concerning subject's financial affairs. The case file is publicly available, but you'll have to go to Washington to examine it (or else hire a Washington researcher). To interpret it, you may need an accountant's help.

If subject is involved in civil litigation and his or her personal finances become relevant, federal and state tax returns may be produced during pretrial discovery and possibly entered in the case file; they may also be entered as evidence at trial. Be on the lookout, especially, for divorce cases involving alimony, child support, child custody, or property division; tax returns are often key evidence in such cases.

If subject is indicted on white-collar criminal charges such as insider trading or income tax fraud, the prosecution may enter federal and state tax returns and other personal financial documents as evidence.

# 8.13 Securities and Exchange Commission Filings; State Securities Filings

A publicly held corporation's registration statement, prospectuses, proxies, Form 10-K's (annual reports), and other filings will reveal financial data about the firm's top officers, directors, and beneficial owners of 5 percent or more of the stock. The data revealed will include salaries of top officers, payments to directors, and the amount of stock owned by both insiders and five-percent owners. In various filings, the firm's officers and directors must reveal any personal financial transactions, gifts of stock, etc., that might create a conflict of interest.

In addition, SEC filings provide information about holdings in the firm by members of an officer's or director's immediate family and holdings by his or her family's trusts, estates, or foundations.

The SEC requires filings not just from the 11,000 publicly held corporations but also from about 12,000 brokers and about 5,000 management investment companies.

To find out if your subject (or a company in which he or she is involved) has ever been the target of an SEC investigation for violation of securities laws, go to the *Securities Violations Bulletin*, an SEC quarterly publication that is consolidated into volumes. Look in the index (or perform a LEXIS search) for subject's name and the names of relevant companies. If you find anything, write to the SEC for copies of the opening and closing reports of the investigation. If you then want other documents referred to in these reports, you may need to make a Freedom of Information request.

Note that the SEC only requires filings from companies that sell their stock across state lines. Many companies, however, sell their stock only within a single state. These companies file statements with the state securities regulator that are similar to those required by the SEC. Like SEC filings, the state filings are public documents.

# 8.14 Disclosure/Spectrum Ownership Database

An excellent investigative tool for tracing the financial affairs of important businesspeople is the Disclosure/Spectrum database available from DIALOG. Disclosure/Spectrum provides full-text searches of the SEC filings of over 5,500 companies. This means you can type in subject's name, and Disclosure/Spectrum will tell you every instance in which he or she has been reported as a five percent or more stockholder, an insider stockholder, a director, or an officer of any of the companies in the database.

## 8.15 Financial Data on Civil Servants, Elected Officials, and Government Appointees

Salary information is available on millions of Americans working at city, county, state, school district, or federal jobs. On the local and state level, there may be a salary roster giving each employee's base pay, overtime pay, and total pay for the year. If not, the job title of a civil servant, and the salary range for that title, will surely be on the public record. On the federal level, the Freedom of Information Act entitles you to be told the job title, grade, salary, and duty station of most civilian federal employees; contact the employing department or agency. For information on former federal employees, write: OPF/EMF Access Unit, P.O. Box 18673, St. Louis, MO 63118.

All elected officials on the federal level and most elected officials on the state and local level must file periodic financial disclosure statements.

Approximately 10,000 federal political appointees, ranking civil servants, and congressional aides must also file disclosure statements. Laws regarding financial disclosure by state and local legislative aides and executive appointees vary from locality to locality (in New York City, for instance, the city clerk's office has financial disclosure statements of all city employees earning over $30,000 per year), but in general the salaries of such people are a matter of public record.

For salary and expenses of members of the U.S. House of Representatives, their staff members, committee staffers, and other House employees, see the quarterly *Report of the Clerk of the House*. For similar information on U.S. Senators and Senate staffers and employees, see the biannual *Report of the Secretary of the Senate*. For salary information on approximately 3,000 top federal appointees, see *U.S. Policy and Supporting Positions* (the so-called Plum Book).

## 8.16 Financial Data on Candidates for Public Office

Every candidate for federal office and most candidates for state or local office must file campaign financing reports with a designated agency. On the federal level, the statements are kept by the Federal Election Commission; on the state and local level, by state and local boards of elections. This holds for candidates in party primaries as well as general election candidates. Over one million Americans have filed such reports while pursuing public office over the years.

## 8.17 Government Statistics and Other Indirect Indicators

A vociferous local Klansman drives a bulldozer for a construction company. You get a tip that he's also involved in an extortion racket. You may lack the resources to look into this allegation fully, but you can at least make a rough estimate of what subject's legitimate income is—to see if he's living beyond his means. However, you know nothing about the construction business and its pay rates.

The U.S. Labor Department's Bureau of Labor Statistics issues reports that tell the average earnings in hundreds of occupations and trades broken down by area of the country and type of firm. The bureau's annual multi-volume *Area Wage Survey* and its monthly *Employment and Earning* are widely available in public libraries. In addition, you may be able to get more detailed breakdowns from bureau databases; contact its press office at (202) 523-1913. For still more detailed information, contact your state labor information office. State labor departments often collect information more detailed than that provided by the federal government, tracking as many as 1,000 occupations and breaking down statistics to the county level. In addition, the state labor department will have on file—and compile statistics based on—the quarterly unemployment contribution reports from every employer in every county. (Note that none of the information you receive from either federal or state labor departments will be personal information about individuals.)

The federal or state Labor Department also will have copies on file of union contracts, which stipulate wages, linkage of wages to years of service, overtime rates, hours, etc.

A phone call, by a person allegedly seeking work, to the personnel office of subject's company, to the union hall, or to a former employee of the company—with a general question about wages for the job category in question, annual increases, etc.—may elicit further information.

Statistical income estimates are available from other sources for a wide range of job categories. For instance, the U.S. Education Department's annual Higher Education General Information Survey (HEGIS) compiles information on professorial salaries by rank and contract length at individual colleges and universities. The reports provided by each school can be obtained from the National Center for Educational Statistics, the state board of higher education, or the school itself. Other excellent sources are the nation's professional and trade organizations, many of which publish detailed salary and benefit data according to job category, region, and other indices, e.g., the American Payroll Association's annual "Survey of Salaries and the Payroll Profession."

A knowledge of the statistics on local wages, salaries, and fringe benefits may help you avoid hasty assumptions. For example, attorneys are generally regarded as a high-income group, especially by fans of "L.A. Law."

Yet many attorneys in your region (especially those working as public defenders or assistant D.A.'s) may earn less money than do unionized blue-collar workers in certain job categories.

Be aware, in using wage statistics, that most people have other sources of income besides their primary jobs, and wages from primary jobs often may fluctuate widely. Millions of people work overtime at irregular, unpredictable intervals, thus boosting their annual income significantly. Millions more work at jobs that involve seasonal layoffs. Others work two jobs, often keeping this fact secret from one or both of their employers. Still others may do part-time moonlighting (e.g., police officers who double as private security guards). A person working at a moderate-income job may have sizeable investments as a result of savings through the years, an inheritance, or a dead spouse's life insurance. In addition, there is the underground cash economy: The bulldozer driver (see above) may be getting his extra income not from extortion but from growing marijuana on the family farm. Or the rumor about his illegal activity may be sheer slander emanating from a rival Klan faction: his family's extra income may come from weekend yard sales. In speculating about someone's income, take nothing for granted; always assume that there's something you don't yet know.

## 8.18 Bank Account Records

A bank will not tell you how much money subject has in his or her account or any other details without subject's written permission. However, someone with a clear need to know—such as a potential employer or landlord—may be told whether or not the person has an account at that branch and whether or not the account is "satisfactory" (i.e., is not overdrawn).

## 8.19 Loan, Job, and Apartment Applications

In researching a subject's background assiduously, you may from time to time happen upon a copy of a loan, job, or apartment rental application filled out by subject in the recent past. Such a document may provide some interesting details about subject's salary history and other financial matters. Be aware, however, that applicant will have slanted the information to present himself or herself in the most favorable light.

## 8.20 Credit Card Information

Records of credit card transactions are confidential. However, if your subject is a government employee, you can make a Freedom of Information request for access to the records regarding any government credit card he or she has used.

## 8.21 Financial Information on Nonprofit Employees and Consultants

Federal 990 forms filed with the IRS by charities and other tax-exempt organizations may contain salary and expense disbursement information about the organization's top officers, and also the amounts paid to outside counsel and consultants. If the organization is the recipient of federal grants, it must file additional information about the finances of officers and consultants with the agency or agencies providing the grants.

You should also check the tax-exempt organization's annual filings with the state's department of state.

## 8.22 Financial Information on Labor Union Officials and Employees

Forms LM-2 and LM-3, filed by union locals and higher union bodies with the U.S. Department of Labor, will tell you the salaries and other disbursements paid to each union officer. It will also tell you this information for each union employee who received more than $10,000 in the past year. Furthermore, it will list all direct or indirect loans of more than $250 made by the union to any union officer, employee, or member, or to any other person or business.

Form LM-30 is a detailed financial disclosure form that union officers and employees must submit if they or their spouses or children have business dealings with a firm whose employees are represented by the union in question. It may include important information about a union official's outside sources of income, his or her investments and debts, etc.

# 9.

## Court Records

### 9.1 Introducing the Court System

Researching court records is a complicated matter because of the wide variety of courts and records systems. If you are starting out as an investigative journalist, one of your first steps should be to familiarize yourself with each court located in or having jurisdiction over your city, county, and greater metropolitan area, and learn how to use the various court indexes to maximum effect.

The trial courts in any locality are divided into local courts and district state courts on the one hand, and federal courts on the other. Each has civil and criminal divisions and a system of higher courts to which appeals are made.

The lowest rung of a state court system is the village, town, city, or county court, which handles relatively minor matters—say, claims of up to $10,000, and misdemeanors like prostitution and driving while intoxicated. These local courts, which usually do not offer jury trials, go by many names—magistrate's court, district court, city court, or superior court. Some may have specialized jurisdictions, e.g., small claims court (to handle claims of $1,000 or less), traffic court, or landlord-tenant court. These courts exist throughout every state, although their names and functions may vary from town to town within a single state according to local custom.

The next level is the state courts, which offer jury trials of larger monetary claims (say, claims of over $10,000) and the more serious criminal offenses. Anyone disputing a city court judge's decision will also come here for a trial de novo (new trial). These state courts, each of which may have jurisdiction over one or more counties, are referred to, variously, as the district court, superior court, county court, circuit court, or state supreme court (note the overlap in terminology with the municipal courts). Various specialized divisions are usually included in this level of the court system; for instance, the probate court (also known as surrogate's court), which handles the probate of wills, administration of estates, and appointment of guardians; and family court, which handles child custody cases and various other marital and family disputes (and, in some localities, juvenile criminal cases).

The federal court of original jurisdiction is the U.S. District Court, which may cover an entire state or several counties within a state, e.g., U.S. District Court for the Eastern District of New York (Brooklyn, Queens, Staten Island, and all of Long Island). The U.S. District Court handles suits brought by or against the federal government and other civil cases involving federal law. It is also the trial court for all federal criminal violations. Its bankruptcy division handles all bankruptcies of individuals and businesses within its jurisdiction.

## 9.2 Working with Court Indexes

Each court keeps indexes of all civil and criminal cases brought before it. In some courts, the indexes may be searchable at a computer terminal. In others, you may have to look through bound computer printout volumes. In still others, the indexes may be on microfiche, in an automatic rotating card file, or even (for older local records) entered in ledger books. Criminal court cases are listed by defendant; bankruptcy cases, by petitioner; probate actions, by the decedent's name. Civil suits are listed alphabetically by either plaintiff or defendant, although some courts have both a plaintiff/ defendant and a defendant/plaintiff index (or else one index that merges both in a single alphabetical listing in which each case appears at least twice—this is called a "party index"). In general, the plaintiff/defendant or defendant/plaintiff index will list only the first-listed plaintiff and first-listed defendant in multiparty cases.

Some computerized indexes (and some microfiche indexes generated from the printouts of computerized indexes) will list all parties, e.g., the COURTRAN system, used by several federal district courts, which combines index and docket in a single database. Such all-party indexes are not yet the norm in state courts, and even where they exist the coverage doesn't extemd back very many years. If you want the names of those other ten plaintiffs in the 1981 product liability suit *Barnett et al. v. Ace Toys*, you will have to go to the docket sheet or the case file.

To find all the local civil suits involving your subject requires considerable ingenuity if there is no cross-index and/or if indexing is restricted to the first-listed parties. As noted above, the computer terminals at the court clerk's office (if indeed they have been installed yet at your local courts) will probably not be able to search the older indexes, thus excluding the closed or dormant cases from previous years that may be the most important for your purposes.

The following are some methods you might find useful in getting around the limitations of your local court indexing systems. These methods are intended for a worst-case scenario: an index in which cases are listed alphabetically by first-named plaintiff with no cross-indexing and where your main concern is to find cases in which your subject is listed as a defendant or unindexed plaintiff.

- Look up all cases in which the state or city is the plaintiff. Depending on local practice, these may be listed under "People of—," "State of—," or "City of—," or possibly under the official designation of the state or city attorney (e.g., "Anystate Attorney General," "Anystate Department of Law"). Under the most commonly used of these government plaintiff headings, you may find hundreds of defendants listed in alphabetical order.

- Flip through the plaintiff indexes, looking for examples of nongovernment or quasigovernment plaintiffs who file massive numbers of cases. In New York City Civil Court and New York State Supreme Court, hundreds of defendants may be listed alphabetically under the plaintiff headings for the city hospital corporation and certain finance companies.

- Go to the judgment books (see 8.4). If a private plaintiff or governmental entity has obtained a judgment against your subject in the local courts, the case file number will be listed alongside the person's name.

- Whenever you find subject listed as one of several plaintiffs or defendants, check the names of the other parties separately in the index to find other cases that might include subject as a party.

- Whenever you find a case involving subject, look in the case file for references to other lawsuits. Especially look for countersuits or crosssuits. Examine the transcript, if any, of subject's deposition or court testimony to see if he or she was asked about previous litigation. Also, check directly with the other parties (or their attorneys); they may be aware of cases you would otherwise miss.

- Look under subject's name (or that of his or her business) in newspaper databases, indexes, or clippings files for reports of any lawsuits.

- Check for "bad blood" suits. If you learn from the plaintiff index that your subject is suing Mr. X., check to see if Mr. X has independently sued your subject about another matter.

- Use the federal court party index to find possible leads to state court cases indexed only by plaintiff. A person who is suing your subject in federal court might have sued him or her on a related or entirely separate matter in state court.

- Follow clues that emerge from other aspects of your investigation. For instance, check in the plaintiff index under the names of subject's exspouses or ex-lovers, business or political rivals, former business partners (especially in businesses that went bankrupt), former friends or political allies who have turned into enemies, etc. They may be suing subject or current associates of subject. Any time you find an antagonist or associate of subject listed as a first-named plaintiff or first-named defendant, check the docket or case file to see if subject is one of the unindexed parties. (All this should apply to any businesses linked to subject as well as to individuals.)

- Check the bankruptcy court index to see if subject or his or her business has ever filed for bankruptcy. If so, subject or the company will have filed with the court a list of all outstanding suits, in any jurisdiction, to which subject or company was a party at the time of filing, as well as a list of all outstanding judgments against subject or company in any jurisdiction (these judgments, of course, will lead you to still more litigation).

Note that when you are searching court indexes that are not online (e.g., microform, CD-ROM, or bound computer printout volumes), you should always search any supplements produced since the last cumulation.

## 9.3 Finding Cases in Other Jurisdictions

Your subject may have moved around a lot, or may do business in many states simultaneously. How can you identify cases involving him or her in localities other than your own?

Begin by searching for subject's name in LEXIS. This giant electronic law library covers most of the decisions handed down by the nation's federal and state appeals courts since the last century. It also covers trial court decisions in New York and Ohio dating back many years, and decisions in special federal courts such as the U.S. Tax Court.

Author Steve Weinberg used LEXIS and its competitor, WESTLAW, in researching his unauthorized Armand Hammer biography. He located about 300 cases involving Hammer and/or close associates and relatives of the oil tycoon. Reading over the judges' decisions, Weinberg selected certain cases that seemed to warrant further digging. He then went to the various courthouses or Federal Records Centers to examine the case files and photocopy selected portions.

The cases searched via LEXIS will represent only a tiny fraction of those filed nationally each year, most of which either don't go to trial or don't get appealed. However, a case that does get appealed is often a richly complicated one, and the case file may be full of juicy revelations.

As an alternative to using LEXIS, you can search the case law digests in your county bar association's law library. For federal court cases, see *West's Federal Practice Digest*. For state cases, see the various state digests. For New York State, one would go to *West's New York Digest 4d* (for cases since 1978), *West's New York Digest 3d* (cases from 1961-1978), and *Abbott New York Digest 2d* (cases before 1961). Both *West's* and *Abbott* have plaintiff/defendant and defendant/plaintiff tables. Note that case law digests only give comprehensive coverage of appeals court decisions. Trial court rulings may be covered, however, if they involve significant points of law.

Another way to find cases in other parts of the country is to conduct a search of newspaper and periodical databases, including those that cover business and legal publications. A case with general news interest will usually be reported in the daily press; cases with specialized news interest may be reported in regional or local business or trade publications. A newspaper/periodicals search often will turn up several cases not found in LEXIS. For instance, a divorce case involving a certain movie star may not be included in LEXIS, but you may find a hundred references to it in NEXIS or VU/TEXT.

Next, check if the federal district court in your locality has the indexes for other federal courts in the region. In the U.S. District Court for Massachusetts, for instance, you can search the indexes for all New England federal districts. At the time of writing of this manual, LEXIS did not offer nationwide searches of federal court indexes; by the time you read this, such a database may be available. But be aware that federal court cases are only a tiny fraction of the total number of cases filed on local, state, and federal levels.

If you know the cities and states that subject has lived in during a life of much moving around, you will want to check the indexes in each. You will thus find useful *The Guide to Background Investigations* (see bibliography), which tells how to order a records search from each of the nation's federal district courts. The manual includes phone numbers, addresses, search fees, and turnaround time. It also tells in which courts the records are available by phone as well as mail, and the cutoff date after which old files are sent to the nearest Federal Records Center.

## 9.4 Obtaining the Civil Case File

In many courts, there will be separate civil indexes for individuals and corporations. The index will tell you the date the case was filed, the name of the first-listed plaintiff, the name of the first-listed defendant, and a docket number. It will also tell you if the case is still pending or if it is closed. You then fill out a requisition form and give it to the clerk in the file room, who will bring you the file.

Always tell the clerk that you want ALL the folders in the case file, including any supplemental folders with depositions in them. If a number of the documents listed in the docket sheet of an open case are not in the case folder(s), they are probably in the judge's chambers. To make an appointment to examine them, call the judge's law clerk.

If the case is closed, the file may be on microfiche or stored in an archive at another location. The files of closed federal court cases are usually sent to the regional Federal Records Center.

If a case is appealed, all or part of the trial court file is sent to the appeals court. Locating the portion you need can get rather complicated (especially if an appeal is on its way from a lower to higher appeals court). The portion of the file at a particular appeals court can be examined there by a news reporter. After the appeals court decision, the file is sent back to the trial court's file room (and from thence to the court archives) unless there is a further appeal. The briefs filed with the appeals court (which sometimes bring out facts not found in the trial court case file) are usually sent to a depository library.

One might think from the above that chasing down court papers can be extremely time consuming. Fortunately, there is often a shortcut. You call up one of subject's opposing parties in the case (or their attorney). The attorney may have the complete file in his or her office, if the case is still open. If the case is closed, the attorney may still have the file, or the client may have taken it home. Note that this file will frequently have deposition transcripts and documents produced under discovery that are missing from the courthouse file.

## 9.5 Studying the Case File

You may find that a case that began in the late 1970s or early 1980s is still open, yet the folder contains almost nothing except the original affidavit of service, plaintiff's complaint, and defendant's reply. In such instances, the plaintiff may have simply decided not to pursue the case. Or there may have been a delay because of a crowded court calendar or plaintiff's attorney's dilatoriness (civil cases often don't go to trial until several years after the filing). However, if a case is being vigorously pursued, the folders will be filled to overflowing with motions, countermotions, and other legal documents. Some of these will contain useful facts; others will be (for your purposes) legal mumbo jumbo. The case docket sheet is your guide to the case file—it lists in chronological order all papers filed, appearances, process served, orders, etc. Although the docket can be confusing to a nonlawyer, a good rule of thumb is to always examine the original pleadings (the plaintiff's complaint against the defendant, and the defendant's reply) and attached exhibits, any affidavits dealing with the substance of the complaint, and any pretrial discovery materials (especially depositions).

If the trial has already occurred, the transcript (if one was ever made) will probably not be included in the file unless the case was appealed. If you want the transcript, and can't get it from any of the parties, you will have to order it from the court reporter, an expense of thousands of dollars. For this, if no other reason, appeals cases are extremely important for an investigator: The party who is appealing must order a transcript and usually will file it (or significant portions of it) with the appeals court.

## 9.6 Pretrial Discovery

Often, the most valuable material will surface during pretrial discovery— the process by which each side seeks information from the other to clarify issues and strengthen its own case. For an investigative reporter, this material may be more important than the trial transcript itself, because pretrial discovery is much more free-wheeling than a trial in respect to the questions that can be asked and the evidence that can be collected. The reason for this is that pretrial discovery is not limited by the rules of admissibility at trial. All sorts of things come out during pretrial discovery that a jury is not allowed to hear.

Pretrial discovery involves the following procedures that are conducted according to federal or state court rules of civil procedure:

- Discovery and inspection of documents: A party to the action serves process on another party demanding the production of business correspondence and other written materials relevant to the issues at law. In a corporate suit, thousands of pages of correspondence, internal memoranda, and financial records may be produced for inspection and copying. In a libel suit, the plaintiff may demand to inspect and copy the reporter's notes and article drafts, and any documents on which the reporter's allegedly libelous statements were based. Documents produced during discovery will not be found in the court file unless filed in support of a motion or as deposition or trial exhibits (see below). However, if the documents produced are not covered under a permanent protective order you may be able to examine them after the settlement or trial, at the offices of the attorneys for the party that conducted the discovery and inspection.

- The posing and answering of interrogatories: A party to the action serves written questions on an opposing party which the latter must answer in writing under oath. Interrogatories are often used as a means to identify documents that will then be demanded under an order to produce.

- The taking of oral depositions (also known as Examinations Before Trial, or EBTs): In a deposition, a person answers questions under oath from the attorneys for one or more parties in proceedings recorded by a court reporter. The person so questioned (called the deponent) may be a party to the action or a nonparty with knowledge of relevant facts. In the large percentage of civil cases (well over ninety percent) that are settled before trial, the deposition is the closest thing to actual court testimony you will find. Some of the questions posed in a deposition will be designed to gain broad background information (thus, the deponent's attorney may complain on the record that the attorney taking the deposition is on a "fishing expedition"). Depositions can go on for days, generating thousands of transcript pages. Unfortunately for journalists, these transcripts are not always filed with the court, although brief excerpts may be included in support of a motion or brief. The rules of procedure regarding public access to deposition transcripts vary widely in both state and federal jurisdictions. For instance, in some jurisdictions transcripts of depositions and exhibits marked for identification therein must be filed with the clerk of the court for public inspection unless there is a protective order. In other jurisdictions, they are not filed *except* by judicial order.

If the court file does not contain much in the way of pretrial discovery facts, it at least may give you an idea of what types of documents were produced and what lines of questioning were pursued. (Indeed, a motion to compel production of documents may itemize exactly which documents are being demanded.) You can then try to get access from a friendly party. However, you may find that the party who is the target of a particular discovery process has obtained a court order stipulating that access to documents, deposition transcripts, and interrogatory answers will be limited to the parties in the case (or sometimes just to the parties' attorneys) and the judge. On occasion, however, this will only be a temporary order; and the materials will show up in the court file or be easily obtained from one of the parties after the case is settled.

## 9.7 Dealing with Subject's Opponents at Law

The attorneys for your subject's opponent may be your best source for copies of depositions and other pretrial discovery materials as well as the trial transcript. Often the attorneys will let you examine or copy documents if you have information to trade, or if they think you might write an article favorable to their client's viewpoint, or if they're simply curious to see what you might dig up.

Once the appeals are over and the case is closed, the attorney may give the case file to the client, who takes it home and tosses it in a closet. If you want to know about subject's divorce, and your state bars public access to divorce filings, you might approach the ex-spouse to see if he or she will be willing to dig the papers out of the closet. (The case file in a legal battle over division of property, child support, alimony, etc. can be extremely revealing about the finances of one or both parties.)

Even if you don't need their help in obtaining deposition transcripts and other documents, it's still useful to talk with subject's opponents at law. They may give you much valuable information off the record. While backgrounding a certain businessman a few years ago, I found a single case in the court index in which he was being sued for fraud. Meeting with the plaintiff's attorney, I was treated to a fascinating lecture on subject's corporate veil and his ties to organized crime, mostly based on private investigators' reports and confidential internal documents of subject's business entities. I learned more about subject in that one session than during weeks of prior research.

## 9.8 Criminal Court Records

### Arrest and Conviction Records

Criminal records are kept by local county or municipal courts, state trial courts at the county level (which also send copies to a central state repository), and federal trial courts. Each criminal court will have an index listing the defendants alphabetically and telling the charge, the disposition (if any), and the case file number.

Public access to arrest and conviction records varies from county to county as well as state to state. In general, if a person has been convicted of a felony or misdemeanor (or if the case is still pending), this information can be found in the criminal court index, although in some localities you will need a release to obtain a records search. If the person was acquitted or the case was dropped, the record usually will be expunged.

In attempting to find out if someone has a criminal record, there are two basic approaches: to search the records of the state repository, or to search at the county (trial court) level. If you search the state repository (assuming these records are available without a release), you will miss many cases—the trial courts at the county level are often negligemt or tardy in forwarding data to the state repository, and some state repositories do not accept misdemeanor or lesser felony records. On the other hand, if you just search the records in subject's home county, you may get reliable results for that county—but you'll miss subject's felony conviction in the next county. Experts at backgrounding suggest that you always search both the state repository and the trial courts of those counties where subject is most likely to have gotten into trouble.

The Guide to Background Investigations describes the procedures for obtaining a criminal records search from every county and every central state repository in the nation. It tells which jurisdictions will provide information only by mail and which will provide it by phone. The book also gives the separate procedures for obtaining misdemeanor and felony records and tells which jurisdictions require a release. Finally, it includes city/county cross references so you can match any city with the county courthouse that controls its criminal records.

You can also find out from the The Guide to Background Investigations how to obtain a criminal records search of federal district court files. It is not yet possible to search the entire federal court system at once—you must make a separate request to each district. However, there is a national database that keeps track of federal prison inmates. If you call the Inmate Locator Line at (202) 307-3126 and give subject's name, they will tell you if his or her name is in the database, which covers inmates back to 1981. If subject's name is in the system, the locator service will provide you over the phone with conviction, sentencing and parole information as well as telling you where subject is or was incarcerated. (For records prior to 1981 you must make a special request.)

Another shortcut is to do a full-text search for subject's name on LEXIS. If subject ever appealed a conviction to a state or federal appeals court anywhere in the country, LEXIS will probably have the appeals decision. You can search a single state or any combination of states. You might also check the LEXIS Federal Sentencing Library and the LEXIS file that covers all federal Racketeering & Corrupt Organizations Act case law, both criminal and civil.

As noted above, an arrest and prosecution record may be expunged if the person was acquitted or the charges were dropped. In such cases, you will have to go beyond the ordinary criminal court records to find out anything. Here are a few suggestions:

First, try the police or sheriff's department arrest logs and jail books. These records, if available to reporters in your locality, will tell you the offense the person was charged with and the date of arrest. In addition, the jail book will tell if subject made bail and how. These records are sometimes not expunged along with the court record, either because there is no policy of doing so or because of inefficiency. The problem is that the records are filed chronologically, so if you don't know the approximate date of the arrest it will often be difficult to find the entry.

Second, consult local newspaper databases and "morgues." If all police and court records have been sealed or expunged, this may be your only practical way of finding out about subject's arrest, but be aware that it's a hit-and-miss method—especially in cities with high crime rates where a majority of arrests are never mentioned in the press.

Third, test the memory of a longtime crime reporter or someone in law enforcement who specializes in tracking the type of criminal activity in which you believe subject has engaged.

Fourth, ask a friend in law enforcement to run a check on subject in the National Crime Information Commission (NCIC) computer network, which contains data on felony arrests and convictions throughout the country. Be aware, however, that your friend could get sacked for this— reporters are not supposed to have access to the NCIC.

Fifth, check with your state or city's Crime Control Commission, if there is one. It may have a clippings file, and its research director may have a long memory (especially for cases involving organized crime).

Sixth, look for evidence of subject's possible arrest record while searching through civil case indexes and files. Some of the civil cases that may have relevant information are:

- Actions by the state attorney's office seeking an injunction or other civil remedy against subject's alleged illegal activity. Such an action may precede an indictment or follow an unsuccessful indictment.

- Civil fraud suits against subject, including RICO actions filed in federal court against subject and his or her associates either by the government or a private plaintiff. These also may precede criminal indictment.

- Civil rights suits filed by criminal defendants against the police officers who made the arrest and/or the prosecutors.

- Suits filed by crime victims or their families seeking monetary damages from an alleged rapist, hit-and-run driver, etc.

### The Official Case File

This is accessed via the index case file number. Generally it will contain the information sheet and complaint against defendant, arrest and search warrant applications, the indictment, the bail affidavit and receipt, subpoenas for witnesses, motions and answers, the exhibits list, the judge's instructions to the jury, and the verdict and sentence. The file may also include the trial transcript, the transcript of defendant's bail hearing (which may include detailed testimony from police officers as to why they think defendant is too dangerous to set loose), and the transcripts of witness depositions taken before trial.

Public access to the case file will vary from locality to locality and according to the disposition of the case. Generally while a case is current the file can be examined by reporters, although you may have to go through the prosecutor's office to do so. If the case was dropped or the defendant was aquitted, the case file will usually be sealed. If the defendant was convicted, you may need a release to examine the case file.

### Appeals Briefs

If a criminal case is appealed to a higher court, the briefs prepared by prosecution and defense will remain part of the public record no matter if the trial court case file is sealed. These briefs, often fifty pages or more in length, may be your best window on the indictment and the trial. Even if you have access to the trial court case file, you should look at them, since they sometimes bring out facts or viewpoints not contained in the original record. Federal court appeals briefs usually end up in the Library of Congress or a local law library or archive designated as a depository, e.g., the library of the Association of the Bar of the City of New York. Likewise, briefs from state appeals court cases will often be found in a university law school or bar association library in the state in question.

### Subpoena Records

The records of subpoenas issued to witnesses in a Grand Jury investigation or trial may be available depending on the locality. If so, contact the subpoenaed person to ask whether they would be willing to discuss the testimony they gave.

### Prosecutors' and Defense Attorneys' Files

Both sides will usually have extensive files that go beyond anything in the official court file, including much evidence inadmissible at trial. A prosecutor seeking publicity may allow reporters to see portions of this file (e.g., portions of wire-tap transcripts in an organized crime case) or will brief them on what's in it. Defense attorneys may also cooperate with a reporter, especially if defendant was the victim of egregious prosecutorial or police misconduct—the reporter is given inside information as a means of exonerating the defendant in the court of public opinion if nowhere else. Of special interest in the defense team's files will be the reports of private investigators hired to gather evidence for the defense.

### Contacting The Victim

If you find an old newspaper article about subject's arrest but no article about the disposition of his or her case—and the court records are sealed—you can always try to obtain information from the victim. Often the article will give the victim's name and address. The victim or the victim's surviving relatives will probably recall much detail about the case (they may even have portions of the court file), and may have closely tracked the subsequent activities of the defendant. If your state has a victims' rights statute, subject's victim may have obtained copies of sentencing reports and may also have gained access to ongoing information about subject's parole applications and hearings, the conditions of his or her parole, etc.

Don't restrict your inquiries to victims of violent crimes. I have found that the dupes of a white-collar scam can be unremitting in their thirst for revenge against the person who scammed them, even a decade or more afterwards.

### Civil Case Files As A Record Of Criminal Activity

A large percentage of civil cases essentially revolve around plaintiffs seeking civil remedies for what are alleged to be illegal activities, especially fraud. Often the complaint in such a case will present as clear a picture of defendant's criminal mind as any prosecutor could provide (the plaintiff's attorney becomes, in effect, the prosecutor). Thus, if you are unable to get the court file on the prosecution of George Roe for selling forgeries of Renaissance paintings, the civil case filed against him the previous year by the purchaser of one of his so-called old masters will give you a pretty good idea of why Roe decided to plead guilty to a misdemeanor rather than face a jury of his peers on felony charges.

### *When Your Subject Is Not The Defendant*

A criminal case file or trial transcript often becomes important in tracking someone other than the defendant. Generally your attention will be drawn to such a case by an old newspaper article rather than the court index. Your subject, according to the clips, may have been an unindicted co-conspirator, may have testified for the prosecution in exchange for immunity in this or another case, or may simply have been a witness who was questioned at length about his or her past during cross-examination. Indeed, your subject may even have been the purported victim of the crime. In 1981, one Richard Dupont, a former gay lover of attorney Roy Cohn, was prosecuted in New York for harassing Cohn. Dupont's attorney, John Klotz, turned the trial into a probe of Cohn's own criminal activities, providing future researchers and historians with a remarkably detailed portrait of New York's premier power broker. When I interviewed Klotz afterward, he told me about evidence he had not been allowed to present during the trial (and questions he had not been allowed to ask) about Cohn's criminal activities. I also spoke with Dupont, who gave me a wealth of accurate details about Cohn, his lovers, his business partners and clients.

## 9.9 Other Specialized Court Records

For housing court records, see 4.22; for bankruptcy court records, 8.10; for U.S. Tax Court, 8.12.

# 10 ·

# Backgrounding the Individual—Miscellaneous Records and Resources

## 10.1 Vital Records

Vital records include the indexes and certificates for births, marriages, divorces, and deaths. Access to these records differs from state to state, and often county authorities will have a policy differing from the state's.

Issuance of birth and death certificates is handled by the state vital records bureau in most states. However, certificates are also issued by about 7,000 local registrars throughout the country. As for marriage and divorce records, the state vital records bureau may keep a central index of marriages and divorces but usually will refer you to the county or city for a copy of the certificate.

The U.S. Department of Health and Human Services has prepared a pamphlet, "Where to Write for Vital Records: Births, Deaths, Marriages, and Divorces," which gives the address of each state vital records bureau and tells how many years back the state records in each category go (earlier records are found on the county level only). This pamphlet is chiefly intended for people who want certified copies of their own records and already know the date of the given event and other identifying details. Journalists, however, are not usually interested in obtaining a copy of the certificate, only the information on it. The index alone may have what he or she needs (e.g., date of birth and mother's name). If the journalist needs more information, an abstract of the certificate, rather than a certified copy, may be sufficient.

The following steps are suggested to obtain vital statistics about your subject and his or her family as easily as possible. First, if the recorded event(s) occurred in your own county, go directly to the local authorities. Marriage and divorce notices are often filed in ledger books in the county or city clerk's office. Birth or death records may be at the same office or at the county or city health department.

If your county no longer issues birth and death certificates, it still may have the old records or at least the old indexes available for public inspection. In addition, the old indexes may be available on microform at the local public library or the nearest Family History Center maintained by the Church of Jesus Christ of Latter-day Saints (the Mormons).

If the event occurred after the county stopped maintaining records in the given category, or if you don't know in which county the event occurred—or if it occurred in another state—you should call the vital records bureau at your state's (or the other state's) health department. Before requesting any specific record, inquire about the bureau's policy on that category of records: Are copies available to the general public? If not, can you at least get an abstract of the record? Can you get index information—or at least, the date of the event—over the phone?

In about ten "open record" states, there are no restrictions on obtaining vital records at the state level. In some of these states, you can get the information read to you over the phone, and if you want a copy or abstract, you can order it over the phone and pay by credit card. (Elsewhere, you will have to write a letter, enclosing a check or money order.)

Sometimes there are restrictions at the state level on releasing certified copies of birth certificates, because of concerns over ID fraud. This is why you should check if it's possible to get an abstract of the birth record (which could not be used as readily for fraudulent purposes) or at least the information included on the annual index. (If even this is not available from the state, ask the clerk in what county the event occurred—then try the county office.)

In some states copies of birth and death certificates are not provided to anyone except the person to whom the birth record pertains or to members of the deceased's immediate family. But often no proof of identity is required of anyone making a request either in person, by mail, or over the phone. Indeed, persons requesting the information by mail or telephone (as opposed to those appearing in person) usually aren't even asked to fill out an application. In addition, states that strictly limit access to certified copies of vital records may maintain an open index of those records, which will give you the date of the event, birth mother's name, etc.

If the state simply won't help you, a county registrar often will. Although as noted above some states have taken control of vital records out of the county's hands, most of these same states have allowed local registration to continue in selected metropolitan areas.

If you find both local and state records closed to the public, it is possible that they were photocopied by the Mormons prior to the passage of the law restricting access. It is also possible that the county or state government itself has provided copies of the otherwise unavailable birth and death indexes to the local public library's genealogy division or to the state archives.

If you can't find subject's birth record in either the state or county files, this may mean you have been given a false lead about subject's place of birth. It also may mean that a record of the birth was never filed. In such cases, look for the birth information in local church baptismal and christening records, which are almost always open to the public.

Date and place of death is the easiest item to find—just look in the Social Security Death Index (see 4.11) at the local Family History Center or your public library's genealogy division. If the information is not there, you might obtain it from a newspaper obituary or funeral notice, cemetery records, church funeral records, or possibly from the local coroner's office.

A new category of vital records is emerging in cities with ordinances allowing the registration of gay and lesbian domestic partnerships. Several newspapers, including the *Minneapolis Star Tribune*, are now publishing announcements of gay and lesbian partnerships in their wedding and engagement pages. For more tips relating to vital records, see 4.11 and 7.2.

# 10.2 Department of Motor Vehicles Records

Motor vehicle and owner/driver records are easily available. Each state DMV offers searches of most or all of the basic records by mail. Typically a search will cost you about $3 with a turn-around time of no more than two weeks. For quicker service you can access nationwide commercial data nets. These are the work of vendors who purchase the DMV data on magnetic tape, download it, add their own software, and offer searches via information brokers. In addition, PC users who need to search DMV records on a daily basis can, in some states, pay a subscription fee to the DMV and obtain direct online access to certain of its databases.

Abstracts of driver's licenses are available from most states. Depending on the state, the abstract may tell you subject's date of birth, license number (in some states this is the SSN), address, sex, physical characteristics, and driving restrictions (if any).

In many states, you can also obtain subject's driving record without a signed release (see *The Guide to Background Investigations* for state by state instructions). The driving record will tell you where and when subject has been convicted for a moving violation, including drunken driving. It may also include information on accidents and give the location and file number of each accident. Driving-record information is maintained on any individual convicted in a state, whether or not he or she is a resident. Hence, you should try states in which subject has traveled frequently for business or pleasure, as well as his or her home state. Once you obtain an abstract of subject's driving record, you can go on to check the traffic court record on each violation as well as State Police accident reports, records of insurance coverage at the time of the accident, etc.

If a person does not have a driver's license, it may be because the license was revoked or allowed to lapse. Check the DMV's records on lapsed and/or revoked licenses.

The DMV also offers auto tag, vehicle ID number, and vehicle owner searches. Information brokers can perform these searches on an all-state basis. Give them the license plate number, and they'll give you the owner's name and address. Give them the vehicle ID number, and they'll give you the owner's name and address as well as the year, make, and model of the vehicle and the names of any lien holders. Give them subject's name and address, and they'll come up with a list of all registered vehicles he or she owns.

As noted in 4.20, the state DMV may maintain a register of state ID cards issued mostly to people who don't drive or have had their licenses revoked but need valid ID for check-cashing purposes.

The DMV in your state may also handle registration information for boats and recreational vehicles.

## 10.3 Selective Service Records

Under the current selective service registration law, any male born in 1960 or later is required to register at his local post office upon reaching the age of eighteen. The completed registration form, including name, date of birth, SSN, address, and phone number, is sent to the Selective Service Board data center in Illinois, where a number is assigned to each registrant. This number, the first two digits of which are registrant's year of birth, may one day become an important identifier if the draft is reinstituted. At present, however, the registration forms and numbers have little significance, especially since the Selective Service Board does not reveal registrant's address (or any other personal information) except to law enforcement authorities. The only thing the board will tell you is whether or not subject is in compliance with the registration law. To get this information you must write to the board and include the registrant's full name, date of birth, and SSN.

The old registration system, in effect until 1975, produced quite useful records for investigators. The individual registered with his county draft board and was given a number and a draft classification. His classification records were kept at the local board and were open to the public. After the draft was abolished, these records were sent to regional Federal Records Centers around the country, where they are still available to the public. (To obtain them, you must supply the registrant's full name, date of birth, and home address at time of registration.)

The classification records include the individual's selective service number, birth date, every classification (1-A, 4-F, etc.) he ever held and the date the notice of classification was mailed to him, the date of his Armed Forces medical exam and the fact of whether he passed or failed, the date (if ever) that he entered the Armed Forces and whether he enlisted or was drafted, and the date he left the Armed Forces (assuming that he was discharged before the old registration system was abolished). However, there are gaps in these records: The date of leaving the Armed Forces is not always included in the records of World War Two veterans; men who enlisted before they reached the age of eighteen (the time at which draft registration was required) are not included at all, even if they became career soldiers; and Armed Forces women are not included.

In the days of the draft, personnel managers and bank loan officers often found the Selective Service number useful in checking the accuracy of an applicant's statements about his past. This was because the number contained (as does the SSN) coded information. Even today—if you happen across the number in an old job or apartment rental application—it can be useful in finding out where a person grew up. It consists of four groups of digits connected by hyphens: The first group designates the state or territory of the United States where the individual first registered; the second group corresponds to the number of the county draft board within that state or territory; the third group corresponds to the last two digits of the year of registrant's birth; the fourth group is the registrant's local draft board registration number.

The following are the numbers that correspond to each state and territory:

| | |
|---|---|
| 1 Alabama | 25 Nebraska |
| 2 Arizona | 26 Nevada |
| 3 Arkansas | 27 New Hampshire |
| 4 California | 28 New Jersey |
| 5 Colorado | 29 New Mexico |
| 6 Connecticut | 30 New York |
| 7 Delaware | 31 North Carolina |
| 8 Florida | 32 North Dakota |
| 9 Georgia | 33 Ohio |
| 10 Idaho | 34 Oklahoma |
| 11 Illinois | 35 Oregon |
| 12 Indiana | 36 Pennsylvania |
| 13 Iowa | 37 Rhode Island |
| 14 Kansas | 38 South Carolina |
| 15 Kentucky | 39 South Dakota |
| 16 Louisiana | 40 Tennessee |
| 17 Maine | 41 Texas |
| 18 Maryland | 42 Utah |
| 19 Massachusetts | 43 Vermont |
| 20 Michigan | 44 Virginia |
| 21 Minnesota | 45 Washington |
| 22 Mississippi | 46 West Virginia |
| 23 Missouri | 47 Wisconsin |
| 24 Montana | 48 Wyoming |

49 District of Columbia
50 New York City
51 Alaska
52 Hawaii

53 Puerto Rico
54 Virgin Islands
55 Guam
56 Canal Zone

## 10.4 Military Service Records and Discharge Papers

Anyone can obtain anyone else's service record by sending a copy of Standard Form 180 ("Request Pertaining to Military Records") or a letter requesting the information to the appropriate agency. For the records of discharged, deceased, and retired military personnel, send your request to the National Personnel Records Center (Military Personnel Records), 9700 Page Boulevard, St. Louis, MO 63132. Instructions on where to send requests for the records of active-duty personnel as well as reservists and members of the National Guard are contained on the back of Form 180.

On Form 180 or in your letter, state that your request is being made pursuant to the Freedom of Information Act. Ignore the notice on Form 180 that you need to get the veteran's signature on a release authorization (in fact, such a release is only necessary if you are requesting health records or detailed personnel records that are not part of the service record usually released).

You can obtain a copy of Form 180 from the National Personnel Records Center (NPRC) or from a local military base or veterans' organization. When you get the form, make photocopies for future use.

If possible you should provide subject's full name correctly spelled, SSN number, approximate dates of service, and branch of service. If you do not have all this information, the NPRC staff may still be able to find the record.

The NPRC currently holds over 50 million military personnel records. An estimated 18 million records were destroyed in a 1973 fire, including about 80 percent of the records on Army personnel discharged between 1912 and 1959. Although about 1.6 million of these records have been partly reassembled from other sources, no one knows exactly what is missing—there was no index.

The information available from the NPRC under a Freedom of Information request includes date of birth, dates of service, dates of rank/grade changes, awards and decorations, duty assignments, current duty status, civilian and military educational level, marital status, and the names, sex, and age of subject's dependents. Releasable records also include court-martial records if unclassified and a photograph if available. In your letter or in a statement appended to Form 180 you should request each of the above items specifically. Note that the NPRC will NOT provide you with

subject's SSN, address, or telephone number (for instructions on contacting retired, reserve, or active-duty military personnel, see 4.25). The
NPRC also will not release disciplinary, discharge, or medical records
without the veteran's written consent or the written consent of the next-
of-kin of deceased veterans.

If subject's military records were destroyed in the 1973 fire, try to obtain
his military discharge papers. These are available at the county clerk's
office of the county in which subject filed them. The trick is to figure out
which county that might be. Your best bet is subject's home town or the
town in which he ended up when discharged. Note that the Army sends
copies of discharge papers to the Adjutant General of the veteran's home
state.

## 10.5 Telephone Company Records

The telephone company compiles your monthly bill based on computerized records of all local and long-distance numbers called, date and time,
duration, etc. To access these records, you will need an inside source at
the phone company. If you don't have a source of your own, you can go
through a private investigator who has one. Other ways of accessing these
records include calling the billing office and pretending to be the telephone
subscriber, and calling the telephone company's special number for toll
records and pretending to be a billing office employee. (Such tactics are
NOT recommended by this writer.)

If your subject is a government official, you can make a Freedom of
Information request to examine the toll records of all calls made from or
charged to his/her office phone or official car phone, or all calls that were
charged to his/her government-issued telephone credit card. You can then
reverse the numbers and see if subject is calling his or her broker or
engaging in other personal business to an unreasonable extent on the
taxpayer's time and at the taxpayer's expense.

## 10.6 Medical Records

Prior authorization from a patient is required to get medical records from
a hospital or a physician's office. However, bits and pieces of subject's
medical history can sometimes be gleaned from the public record. Subject's
driving restrictions may be included in an abstract of subject's driver's
license (see 10.2). Subject's driving record may include multiple drunk
driving incidents, suggesting a serious alcoholism problem; it also may list
accidents in which subject was involved (the accident report may tell if
subject was seriously injured). Registration records from the draft era may
reveal if subject failed his Armed Forces medical exam and was given a
4-F status, although the specific medical reasons will not be provided.

Income tax returns filed by subject as evidence in a court case may reveal large deductions for medical expenses.

In many states, worker's compensation records can be obtained without a signed release. Typically these records are filed by name of claimant, but you must furnish also the SSN and/or date of birth. In a few states, the records are filed by name of company, in which case you must know where subject has worked in order to access them. *The Guide to Background Investigations* will tell you where to write for these records (and the restrictions on access, if any) in all fifty states.

In searching court indexes, you should keep alert for any case in which subject's medical condition could be an issue (for instance, a product liability suit, a suit against a health insurance company by a policy holder, or a personal injury suit alleging negligence by an auto driver). Frequently in such cases, plaintiff will have to undergo a pre-trial physical or mental examination. The results of this exam may be filed as an exhibit to a pretrial motion or reply; it also may be presented as evidence at trial. Plaintiff may be questioned at length about his or her medical condition in a pretrial deposition and in court testimony. Likewise, the physicians who have examined or treated plaintiff may be questioned in depositions and at trial.

One type of case that is especially rich in medical data is the malpractice suit filed against a doctor or hospital. These cases may be easy to find: Docket Search Network, a Chicago company, offers a Physicians Alert service covering several states; for $10 a subscribing doctor can obtain a list of malpractice suits filed by a given patient.

Further information on medical problems can be gleaned from the county judgment dockets. You may find that several judgments against subject have been obtained by physicians and hospitals because of nonpayment of bills. The judgment will give you the case file number, and you can then retrieve the file from the county or state court file room.

Searches of newspaper databases and clippings morgues may turn up a story about an automobile accident, assault, or other incident resulting in serious injury to subject at some time in the past. House organs at subject's place of employment and church bulletins at subject's place of worship may include information about a serious illness involving hospitalization, and may tell what hospital subject was in (or at least give a phone number which can be reversed to find out which hospital it was). Garbological examinations of subject's trash (see 11.4) may turn up empty prescription bottles as well as medical bills.

Sperry Rand's Medical Information Bureau provides life insurance companies nationwide with medical information on insurance applicants. Its database has over 8.8 million records. A recently published manual for private investigators stated apropos of the Medical Information Bureau: "We are getting into some networks of information that the insurance industry wishes no one but them knew about . . . ."

## 10.7 Welfare Records

These are not open to the public. However, your subject might file copies of welfare checks or various welfare documents in a court proceeding (for instance, a family court case regarding child support payments). Welfare records would definitely be entered as evidence in a welfare fraud prosecution.

## 10.8 Immigration and Naturalization Records

U.S. Immigration and Naturalization Service (INS) records on individual immigrants are usually unavailable to the public. They become part of the public record, however, if filed by either the government or the immigrant in a deportation trial or in a suit launched by the immigrant against the INS in federal district court.

When an immigrant becomes an American citizen, a notice is filed with the county clerk's office in the county where he or she was naturalized. This is public record information. It may include subject's age, address, and occupation when naturalized; date of naturalization; date of arrival in the United States; and former nationality.

## 10.9 Voter Registration Records

These are generally kept by the county or city Board of Elections a/k/a Election Commission a/k/a Registrar of Voters. You can check the enrollment books of registered voters, which list all registered voters in each assembly district and their party affiliation or independent status. The lists are arranged alphabetically by street address. If subject is not in the local crisscross directory, he or she may be listed here. Even if not, subject's spouse may be listed, and also subject's children of voting age. Family members may also be listed in back issues of the enrollment books, which provide yet another means of tracing changes of address as well as changes in the composition of subject's household.

Another key resource is the registration records for individual voters. In some cities you can check both current and back records on microform or at a computer terminal. Depending on the particular records system, you may find out subject's current address, the address where he or she first registered locally (if different), telephone number, SSN, party affiliation, dates of registration and re-registration, occupation, and business address. The records may also tell how many times, and when, subject failed to vote in recent elections (this is especially interesting in backgrounding candidates for public office—you may find that your local "good government" liberal or superpatriotic conservative almost never bothers to vote). Finally, the voter registration records may include a specimen of subject's signature.

## 10.10 Federal Election Commission Records

If subject gave $500 or more to any single candidate for federal office or any single political action committee during any single election between 1981 and 1988, and if subject gave more than $200 under these conditions in any election from 1989 on (and if the contributions were lawfully reported), you can find a record of it via the Federal Election Commission's online Contributor Search System. The FEC charges only $25 an hour for online access, and you can download the data and program it with your own software.

In searching via the FEC database, you can try different variations on subject's name and also all other persons with subject's surname in his or her locality. You cannot do this by street address—the names are only entered into the database by town and zip code (to find the street addresses you have to search the microfiche records at FEC headquarters or order specific microfiche or photocopied records by mail). But if you see that local contributors with the same surname have given money to the same candidate, you can check the telephone white pages or a crisscross directory to see if they are from the same household.

The FEC database lists each contributor's place of work and/or occupation. You can thus use it as a kind of city directory to compile lists of people who worked for the same employer in the same city as subject at various points over the years. If subject is the boss of the company this might be a way to find employees he or she sacked (especially executive level employees). For instance, you see that John Brown was listed as working at Ace Missile Components in Houston in 1982; Brown gave $1,000 to a certain candidate that year, but the next reported donation from Brown comes from another company in New Orleans. You can infer that anyone who gave such a large donation was probably a high-level executive at Ace and probably knows a lot about his former boss. If he was fired and is still angry about it, he might be willing to talk.

The Contributor Search System is also useful in finding out if subject has any extremist connections (see Appendix). Often people with bigoted beliefs are very close-mouthed in the presence of outsiders—you only find out about their ferocious inner life from the FEC filings. For instance, you may discover that the mild-mannered Mr. Deeds who is always ever-so-polite to black and Hispanic customers in his store has donated thousands of dollars in the last three presidential elections to candidates of the super-rightist Thunderbolt Party.

## 10.11  Permits and Licenses

A wide variety of professions, trades, vending operations, and service-type businesses require a license from the city or state government or from a quasi-public commission or agency. You can get a list of the licenses required in your locality from the state and municipal government handbooks, but let's take New York City as an example.

If subject operates a grocery or restaurant he or she must have a permit from the City Department of Health. If subject operates a liquor store or bar or sells alcoholic beverages in his or her restaurant, a license from the State Liquor Authority is required. If subject is a teacher, he or she should have a license from the Board of Education. If subject earns his or her living as a real estate salesperson or broker, barber, hairdresser, cosmetologist, billiard room operator, private investigator, or apartment referral agent—or if subject is a notary public—or sells hearing aids—he or she must have a license from the Division of Licensing Services of the state's Department of State. If subject is a taxi driver, he or she must have a chauffeur's license from the State Department of Motor Vehicles and a hack license from the City Taxi Commission. If subject owns a handgun, he or she must have a permit from the Police Department. If subject owns a rifle or shotgun, he or she must have a permit from the City Firearms Control Board. If subject is a master plumber, he or she must have a license from the City's Master Plumber's License Board; if an electrician, from the City's Electrical License Board. If he or she carries a press card, it will have been issued by the Police Department. If he or she is involved in any of fifty-four different trades or types of business—from auctioneer to weighmaster, from bowling alley operator to street vendor of hotdogs— he or she should have a license from the City Department of Consumer Affairs.

The above list of licenses, permits, and certificates (similar to the requirements in other cities) is only the surface of the iceberg. *The City of New York Official Directory* lists no less than 1,600 licenses required within the city by various local, state, and federal agencies for the conducting of various trades or professions or for the sale of various products or services.

Violations of law by licensees in any regulated area may result in summonses issued by city or state inspectors, or civil or criminal actions initiated by state or city attorneys. In addition, complaints against a licensee may be filed by irate consumers with the City Consumer Affairs Department, the State Attorney General's office, a professional licensing board, or the Better Business Bureau.

## 10.12 Professional Licensing

Professional licensing laws are a confused patchwork from state to state. A recent Congressional study found that although 700 professions require state-level licensing in one or more states, only 20 professions require licensing in all states. In addition, the examinations and other licensing requirements may vary widely in their rigor. Thus, a person whose license has been revoked in Vermont may either move to New Hampshire, which has no licensing requirements, or Connecticut, where the licensing body will overlook the prior revocation (so you should always check the prior state's records). Likewise, a person who cannot pass the licensing exam or meet other licensing requirements in his or her home state may move to a state with an easy exam or no exam at all and/or with loopholes in the other requirements.

By submitting to the licensing procedure a person creates yet another paper trail to follow him or her through life. For instance, consulting *McKinney's Consolidated Laws* we find that a licensed cosmetician in New York State will have filed, at one time or another, an application for a trainee license, an application to take the licensing exam, an application for a permanent license, a physician's certificate stating that applicant is free of disease, a photograph to accompany the license application, a diploma from a licensed training center, a copy of any license received in another state or in a foreign country, and change-of-address notices.

Just how much of such paperwork is open to public inspection varies from state to state; you may have to use the state freedom of information law to get it. At a minimum you will want to know if complaints have been filed against subject and if subject has ever been the target of any disciplinary action such as license suspension or revocation, a reprimand, or a fine. Depending on the profession and the laws of the particular state, complaints may be filed with the State Attorney General's Office, the professional licensing board, or the private association representing that profession or trade on the county or state level; occasionally you may find that complaints have been filed and investigations conducted on all three levels. A complaint to the Attorney General's Office may result in a civil suit by the state or a criminal prosecution. A complaint to the regulatory board may result in an administrative proceeding followed by various appeals (and perhaps a suit by the licensee against the board). A complaint to the local professional organization may result in a more informal investigation, perhaps leading to the licensee's suspension or expulsion from the organization.

# 11 ·

# Backgrounding the Individual—Special Problems and Methods

## 11.1 Backgrounding Subject's Educational Past

You may spot a college B.A., M.A., or Ph.D. degree on subject's wall. Or you may learn about the supposed degree from his or her acquaintances or a biographical sketch in a vanity directory. Verifying the degree is a twofold process: first, find out if the college is a legitimate one; second, find out if subject really graduated. This is an important step in backgrounding: A congressional study has estimated that over 500,000 Americans have obtained false academic credentials from so-called diploma mills.

### Checking Out The College
Make sure you have the name right. Some diploma mills utilize a look-alike name so people will confuse them with legitimate schools; for instance, "Darthmouth College" instead of Dartmouth College, or "Boston City College" instead of Boston College.

Once you have the correct name, check out the school in one of the standard reference guides at your public library: *Lovejoy's College Guide*, *The Right College*, or *American Universities and Colleges*. If it's not in these books, you have good reason to be suspicious.

Next check with the regional accrediting association that covers the state in which the institution is located. In the United States only six regional groups, operating under the umbrella of the Council on Post-Secondary Accreditation, are accepted as having the authority to accredit an entire college or university. These are the Middle States Association of Colleges and Schools, New England Association of Colleges and Schools, North Central Association of Colleges and Schools, Northwest Association of Colleges and Schools, Southern Association of Colleges and Schools, and Western Association of Colleges and Schools.

The situation is different for professional training aimed at achieving certification or licensing in a particular profession. Here each profession has its own accrediting agency or agencies for schools or programs serving that discipline. These entities—about fifty in all—are approved by the Council on Post-Secondary Accreditation and/or the U.S. Department of Education; a list can be found in *Lovejoy's College Guide*.

Professional training programs may operate either as independent institutions (e.g., a college of optometry) or as departments or special schools within a regular college or university (e.g., a university engineering department). In the latter case, there may be differences between the standards of the regional and professional accrediting bodies. On the one hand, the professional accrediting body may fail to accredit the given college department even though the college as a whole is accredited by the regional body. (The student's B.A. degree thus is legitimate for all general purposes, but if he or she majored in psychology and the college's psychology department is not accredited by the American Psychological Association, he or she may experience difficulty in getting into one of the better graduate schools.) On the other hand, the professional agency may accredit the relevant department even though the regional accrediting body has refused to accredit the college as a whole.

In checking the accreditation status of any college or professional program, it is important that you focus on the years of subject's attendance. A school that is fully accredited today may not have been accredited when subject studied there in the early 1970s. Likewise, a small school that has lost its accreditation in recent years because of a loss of income (and hence a decline in the quality of its faculty, library services, etc.) may have offered an adequate education when subject attended it in the late 1950s.

Never accept an obscure school's own claim that it is accredited; diploma mills will lie. And do not accept verification from any regional or professional accrediting bodies except those approved by the Council on Post-Secondary Education and/or the Department of Education. A number of unrecognized accrediting agencies have been set up to provide alternate accrediting for unaccredited schools. Although some may attempt sincerely to impose minimal standards, others are just mail drops for diploma mills.

Note that displaying a degree from an unaccredited school is not necessarily fraudulent, nor is an education received from an unaccredited school necessarily illegitimate in the larger sense. A smart student who works hard at his or her correspondence courses may end up better educated than a student who cruises through an accredited school taking easy courses. As alternate forms of education continue to flourish, some currently unaccredited schools will doubtless gain accreditation in the years ahead or, at least, de facto acceptance in the workaday world.

For an evaluation of a particular unaccredited school, check the latest edition of *Bear's Guide to Non-Traditional College Degrees*, which lists hundreds of accredited and unaccredited schools that offer degrees under unorthodox conditions: correspondence colleges, colleges with minimal residency requirements, colleges offering generous credits for life experience and/or for passing an equivalency exam, colleges offering learning contracts in place of required courses. *Bear's Guide* also lists over 100 diploma mills that sell degrees (including medical and law degrees) through the mail with virtually no work required.

If a suspicious school isn't included in Bear's list of diploma mills, contact the state Education Department and the Better Business Bureau; they may have received complaints about it. Another option is to contact the school itself, pretend that you want to purchase a degree—and see how they respond.

## Checking Out The Degree

Once you have determined that subject's college is legitimate, contact the registrar's office to determine if subject's degree is also for real. No registrar's office will give you details on a former student's academic records without his or her written permission. However, most registrars will confirm if subject ever matriculated and what degree, if any, he or she received and in what year. Note that people who falsely claim college degrees on resumes or job applications will often list a school that they attended but from which they never graduated rather than picking a school at random.

*The Guide to Background Investigations* (see bibliography) will tell you what number to call at each college to obtain confirmation of a student's attendance and degree. It will also tell you what identifying information must be provided (some schools require that you provide the former student's SSN and graduation date), and whether the information can be obtained over the phone without a written request.

## Defining The Years Of Attendance

To trace subject's college career, you first need to know which years he or she was really in attendance. If subject graduated in 1984, it does not necessarily follow that he or she was in attendance for the preceding four years; students often drop out for extended periods. If the registrar's office will not give you the dates of attendance but only the date of graduation, look in the alumni directory. Occasionally, these will list the years of actual attendance, but more often students will be listed as members of the class with which they entered; e.g., "class of 1973," which refers to the year subject would have graduated if he or she had taken an uninterrupted four-year course of study beginning with matriculation as a freshman in 1969. The directory will also tell what degree (if any) subject received and in which year. Thus, you will know to search the college yearbooks and other publications for the years between class year minus four and the graduation year. Note, however, that most alumni directories will not tell you if a student transferred from another school, although if he or she graduated after less than three years of continuous attendance you can assume this is probably the case.

### Student Directories

The back issues of the annual student directory will also indicate which years subject was in attendance. Indeed, student directories (which may be divided into separate volumes for graduate students and undergraduates) can be used like a city directory to trace where subject lived each year, whether on or off campus. In the case of a university divided into separate schools or colleges, the directory back issues may tell in which division subject was enrolled at the beginning of each year, thus revealing any sudden shifts in career goals—e.g., from Engineering one year to Liberal Arts the next.

### Yearbooks

College yearbooks will often contain many details about subject's campus activities and friends. Backfiles of the yearbook will be available at the college library (usually in the campus archives) or at the yearbook editorial office. If the college is not within convenient traveling distance, your public library can borrow the volumes through interlibrary loan.

A typical yearbook will include sections on sports teams, fraternities and sororities, campus publications (newspaper, yearbook, and humor and literary magazine staffs), honorary societies, student government, music and drama, and religious and political clubs. For each organization, team, etc. there will be a group picture, with a caption giving the name of each person.

Look first in the yearbook for subject's senior year. Here you will find his or her class picture together with information regarding degree, major, scholarships, fraternity or sorority membership, and possibly the name of subject's hometown and the high school he or she graduated from. There will also be a listing of subject's campus activities, honors, and affiliations with the year or years of each. You can then go to the appropriate yearbooks, look up each cited activity or organization, and get a list of the other students in each picture. In this way you can rapidly collect the names of those former classmates most likely to remember subject well, e.g., subject's sorority sisters, field hockey teammates, or co-workers on the campus newspaper.

Many yearbooks will have a name index listing each page in which subject's name or picture appears; this can make your search much easier, especially if there is no senior class picture of subject or only a picture with no summary of activities. (Some seniors will have nothing but the degree and major listed—not because they were inactive on campus but simply because they failed to fill out the form sent to them by the yearbook staff. In such cases, simply look in the name index for each yearbook during subject's years of attendance.)

### Faculty Directory

Back issues of the faculty directory will give you the names and departments of professors who were teaching at the school during subject's years of attendance. The teachers most likely to remember subject will be those in his or her departmental major. Also, if subject was on a sports team, the coach may remember him or her quite well. If a professor or coach you need to contact has retired or moved to another school, get their new address from their former department or look in the latest edition and supplements of the *National Faculty Directory*, the *Faculty White Pages*, or the *Faculty Directory of Higher Education*.

### Campus Newspapers and Periodicals

The campus newspaper for the relevant years may have published articles regarding subject and/or articles, letters to the editor, or Op-Ed pieces by him or her. Larger colleges have daily newspapers that cover campus life quite thoroughly. In backgrounding several followers of Lyndon LaRouche, I found much information in Columbia University's *Daily Spectator* on their activities as campus protest leaders in the 1960s.

Campus papers are usually available on microform in the college archives; the microform for the years you need can be borrowed via interlibrary loan.

The staff of the college archives may have compiled a newspaper index or clippings files. These are likely to be organized by subject only, but you can zero in on articles mentioning a particular person if you already know a bit about his or her college career from the yearbooks or conversations with former classmates. Even without a clippings file you might be able to find some things quickly on microform, e.g., if you know the person was on the basketball team, just scan the sports pages for the basketball season.

If I were backgrounding the college career of a celebrity, I would find out who the editors of the campus paper were during his or her years of attendance; they will be more likely to remember any newsworthy events regarding subject that most other students would. Also, some of these former editors will have become professional journalists and thus will be naturally sympathetic to a fellow journalist or a biographer.

Note that campuses and the surrounding cultural milieu give rise to a great variety of newspapers and periodicals that are dutifully cataloged by the archives staff. Subject's campus may have two rival dailies; it may also have Jewish, black, or gay/lesbian weeklies as well as weeklies for the law school, engineering school, evening students, etc. There may be a conservative organ, e.g., the *Dartmouth Review* or the *Cornell Review*, and an environmentalist newsletter. There may once have been a counterculture paper. Usually there is a literary magazine and sometimes a humor magazine.

## Checking Out Subject's Thesis Or Dissertation

All universities will have in their library stacks the M.A. theses and Ph.D. dissertations of graduate students who obtained degrees at that school. (Note: not all master's degree programs require a thesis.) Some college libraries will also keep copies of senior honors papers or senior theses. A few colleges require a thesis from every senior.

Theses and dissertations have an Acknowledgements page in which the author thanks his or her adviser, the members of the dissertation committee, other professors, friends, parents, spouse and children, the department secretary, etc. Often twenty or more names will be mentioned. The acknowledgements are especially valuable in finding out the names of ex-spouses and ex-lovers. Also look for any mention in the Acknowledgements of grants or scholarships received by subject.

If you cannot visit subject's college library, you can still easily check out his or her Ph.D. dissertation. Look in *American Doctoral Dissertations* and the *Comprehensive Dissertation Index* at your local research library. These indexes will indicate if your subject indeed has the Ph.D. he or she claims to have.

Next, go to the bound volumes of *Dissertation Abstracts International*, which contain 600-word abstracts of Ph.D. dissertations accepted by universities in the United States and many foreign countries. The abstract is usually written by the dissertation's author. Many master's theses are also included.

DIALOG subscribers can use Dissertation Abstracts Online, which gives the subject, title, and author of almost every American dissertation accepted at an accredited school since 1861 as well as many foreign dissertations. It also includes abstracts of most dissertations since 1980 and of master's theses since 1988. Many research libraries will have this database on CD-ROM.

If you need the entire dissertation, you can order the microform or a photocopy from University Microfilms International, 300 North Zeeb Road, Ann Arbor, MI 48106. If a particular dissertation or thesis is unavailable from UMI, you can get it via interlibrary loan from the school that awarded the degree.

## High School Background

Subject's high school years can be researched just like the college ones. High school yearbooks can often be found at the local public library or the county historical society; if not, a serious researcher can usually get access to copies at the high school itself. The same goes for the weekly high school newspaper and other student publications.

## 11.2 Backgrounding an Individual Via Archival Collections

Although mostly used by scholars, archival materials can be invaluable to an investigative reporter. Such materials may include unpublished manuscripts, scrapbooks, newspaper clippings, leaflets, court documents, self-published pamphlets, personal correspondence, official papers and correspondence of an elected official, internal documents of a political party, internal memoranda of a corporation, oral history tapes and transcripts, photographs, home movies, and video or audio cassettes of TV and radio appearances.

Subject may have donated his or her personal papers to a library either for the sake of posterity or as a tax write-off. A person doesn't have to be famous to do this. Some libraries actively seek the papers of ordinary people for their sociological value. They also seek the papers of people who, although not well known to the public, played an important behind-the-scenes role in (or were well-placed observers of) significant historical events.

Although the chances that your subject has donated his or her papers to an archive are relatively small, there is a much greater chance that papers concerning subject, or memos or letters by or to subject, are contained in the archival papers of a notable person with whom subject was once associated. If I were backgrounding a longtime business agent of an Ohio Teamster local, I would want to know where the papers of the late Ohio Teamster leader and international president Jackie Presser are kept. (Conversely, if I were Presser and still alive, I would be very interested to know that an investigative reporter who once dogged me had donated his files to the University of Missouri.) If I were backgrounding someone who once worked in a minor appointive post in the Carter White House, I would check the indexes of the Carter Presidential Library. If I were backgrounding a longtime local Democratic county chairman, I would go to the state university library and look at the papers of the late governor whose rise to power was engineered by that county chairman.

Let's say you are investigating Tim, the former radical who is now a neoconservative pundit. You know that Tim was once a disciple and top aide to Max Swift, the Trotskyist leader. You check the RLIN database (see below) and discover that Max, who died several years ago, had donated his personal papers and correspondence to the radical history collection at New York University. You also ascertain from the database that a finder's aid has been prepared for the collection describing each box and folder and itemizing the most important individual documents in each. You obtain a copy of the finder's aid through interlibrary loan and learn that the collection includes two folders of correspondence between Max and Tim as well as a draft of a never-published article by Max denouncing Tim after their quarrel. You go to New York City and examine the collection, finding that Max's unpublished article is a bitter tirade which, among other things, accuses Tim of stealing money from the movement and compulsive philandering.

### Archival Databases and Other Guides

To find archival material nationwide, there are two major databases: the Research Libraries Information System (RLIN) and Online Computer Library Center (OCLC). One or both can be accessed via computer terminals at any research library. You type in a person's or organization's name and see how many "hits" you get. Unfortunately, the databases only include relatively brief descriptions of any given archival collection (although individual documents or letters of special importance may be noted), and if subject's name is not in the description you will not get a hit even if the particular collection contains a large amount of information about subject. To find material beyond what's listed under subject's name, search under the names of his or her better-known colleagues or under the relevant organizational titles. In the case of Tim, you would search for material under "Socialist Vanguard Party" or "Max Swift" or "Olga Strong" (Max's wife). You would also search under topic key words such as "Trotskyism" or "U.S. Left." Next you would call each library listed as having significant collections on Swift, Trotskyism, etc., and ask for the archivist who best knows the collection and/or who prepared the finder's guide; he or she will probably recall if the collection contains significant material on Tim. You might also contact the eccentric former "Swiftie" who donated his files to NYU: first, because he might have personal recollections of Tim; second, because he will know his own collection better than any librarian; third, because he might have boxloads of additional documents in his basement.

An alternative to RLIN and OCLC is to look in the Library of Congress' *National Union Catalog of Manuscript Collections* (NUCMC) and the *Index to Personal Names in the National Union Catalog of Manuscript Collections* (the latter includes about 200,000 personal and family names). Unfortunately it is often a decade or more before collections from local libraries get listed in *NUCMC*.

To find material that is not listed either in *NUCMC* or in the databases, look in the subject index in the *Directory of Archives and Manuscript Repositories in the United States* and also in *Subject Collections* and the *Directory of Special Libraries and Information Centers*. Make a list of the repositories most likely to have material pertaining to your subject, and call the archivist at each.

### Government Archives

For government archival collections, see the *Guide to the National Archives*, which describes records held by the various departments and agencies of the federal government. The microfiche "National Inventory of Documentary Sources in the United States: Federal Records" lists about

1400 finding guides to National Archives and Smithsonian Institution collections, as well as to the seven Presidential libraries.

Staff members at the various National Archives collections and Presidential libraries will search for what you need and send you photocopies for a small fee.

### Oral History

Well over a million Americans from all walks of life have been interviewed as part of oral history projects. There are thousands of these projects sponsored by universities; by county, city, or small-town historical societies; and by corporations, trade unions, or churches desiring a record of their organizational life. Participants in these projects include the most unlikely people: For instance, the late Jackie Presser (see above) taped his recollections for a University of Nevada project.

Oral history transcripts and/or tapes (usually audio tapes but occasionally video) are available at the library of the institution sponsoring the project. Well-funded university projects usually have transcripts, but your local county historical society probably can't afford this—you'll have to listen to the tapes and take notes. The purposes of oral history projects vary widely. If a person is only interviewed about a certain peripheral aspect of his or her life, the transcript may run less than fifty pages. If a person is asked to give an account of his or her entire life, the transcript may fill several volumes.

Finding tapes on a particular person is not very different from finding other archival material. Let's return to the case of Tim. You can search RLIN and OCLC for any tapes by Tim or by any of his former comrades (including Max and Olga) or any of his current right-wing associates who might have reminisced about him. You can also search by key word for tapes regarding the history of the U.S. Left, Trotskyism, the Socialist Vanguard Party, and the history of the college campus where Tim had his moment of glory in 1968. . . .

Although such a search will tell you if material relevant to your research has been entered into RLIN or OCLC by archivists at libraries connected to the system, it will not help you to find tapes at small-town historical societies that are outside the system. To find the most promising of the latter collections, see the *Directory of Oral History Collections*.

## 11.3 Subject's Published Writings

The majority of books are written by part-time authors who make their living either in related writing fields such as journalism or who write to communicate their professional findings in science or scholarship or simply for the love of writing. At any moment, hundreds of thousands of Americans are churning out fiction, nonfiction, or verse manuscripts, and these hopefuls may come from any walk of life. Most of them will receive only rejection slips from America's major publishing houses. But many would-be authors, undaunted, will turn to the thousands of small publishers (including vanity publishers). Others will self-publish their own books and pamphlets using home computers and desktop publishing software. Meanwhile vast numbers of amateur and professional writers each year will publish articles, research studies, book reviews, short stories, poems, or letters to the editor in hundreds of thousands of publications ranging from nationally known newspapers and magazines through the most obscure and unindexed church newsletters, sci-fi fanzines, or high school literary magazines.

Never assume that your subject is not among the millions of Americans who have been published in one form or another at some time or another. As an experiment, I asked a friend to check with her family members. She came up with a list of over a dozen siblings, aunts, uncles, and cousins who had written for publication. Their output included bible lessons for a religious newspaper, a college textbook in education, a work on family genealogy, recipes for a cookbook, a self-published autobiography, and in the case of one uncle, communist pamphlets in the 1940s.

### Finding an Author's Books

How do you track down subject's books, including out-of-print, small press, and self-published books? One quick way is to look in biographical dictionary entries about subject. A large proportion of book authors and other writers find their way into such dictionaries. This is true even of rank amateurs: the very energy that propels them to finish a book may also have propelled them into prominence in business or some other field. And however peripheral their writing is to their main career, they will proudly list (in the questionnaire they fill out for the dictionary editors) even that unreadable vanity-press autobiography distributed to only a few dozen friends and relatives.

Serious authors are just about the most exhaustively covered category in the biographical dictionary business. The chief publisher of biographical reference works on writers, with fairly comprehensive listings of each writer's published titles, is Gale Research Inc. Its ninety-volume (so far) *Dictionary of Literary Biography Series* contains biographical-critical entries on even very obscure American writers in a wide range of categories: novelists, poets, dramatists, screenwriters, short story writers, ethnic writers, magazine and newspaper journalists, children's writers, literary scholars and critics, humorists, science fiction writers—the series even includes a volume on beatniks. Gale also publishes *Contemporary Authors* (129 volumes so far) and *Contemporary Authors New Revision Series* (30 volumes so far). All the above reference works and hundreds of others are indexed in Gale's *BGMI* (see 5.2), which includes 845,000 citations to biographical entries on 400,000 authors (these figures do not include the hundreds of thousands of people in *BGMI* who have done occasional writing while following a career in some other field, but whose writings may, as noted above, be mentioned in their entries).

For a list of subject's works currently in print, check Bowker's *Books in Print*, which includes entries for over 860,000 books—virtually every English language book in print from every press, in every genre, on every subject. This eight-volume set, available at most book stores and libraries, includes author and title indexes. It also includes a volume listing nearly 110,000 titles declared out of print or out of stock indefinitely during the previous year. Do not neglect the latest *Books in Print Supplement*, which includes new and forthcoming titles not in the latest *Books in Print* set; and *Forthcoming Books*, which anticipates thousands of titles up to five months in advance of publication. For a more thorough search, try Books in Print Online, which includes complete records for almost 1.5 million titles including over 400,000 declared out of print since 1979 (the database is searchable by author); or Books Out-of-Print, a CD package covering the same out-of-print titles. You can then go to H.W. Wilson's *Cumulative Book Index* (*CBI*) which has annual volumes dating back to 1969 and multiyear volumes back to 1929. *CBI* is available online and in CD-ROM for easy searching of any author's name.

You can also search LC MARC, the commercial version of the Library of Congress' computerized catalog (available from DIALOG). LC MARC contains bibliographic records for every book in English cataloged by the Library of Congress since 1968 (books in other languages are covered beginning at various points in the 1970s). Updated weekly and containing almost three million records, LC MARC is fully searchable by author, title, subject, etc. Library of Congress records prior to 1968 can be searched via REMARC, which is also on DIALOG.

Through the above resources you will probably find the titles of any book by your subject if it was copyrighted and if subject was listed as the author. To find really obscure uncopyrighted books or pamphlets, the most obvious tactic is to just call up subject and ask him for a list. Ned the Nazi may be the most paranoid person in the state of Arkansas, but if you call him at his rural bunker and say you're interested in his writings, he'll probably react like any other proud amateur author—talk your ear off and then send you a complete collection of his out-of-print scurrilous pamphlets by Federal Express at his own expense.

If, however, Ned refuses to discuss his writings with you, run his name through RLIN or OCLC: Pamphlets bearing his name may have been cataloged by a university archive of right-wing nativist publications.

### What To Look For In Subject's Writings
Whatever their literary or scholarly merit, subject's writings may furnish personal background information unavailable from any other source and thus justify the often arduous search to find them. This is most clearly the case when subject has written his or her autobiography or memoirs. Once, after spending weeks collecting information on a New York businessman,

I discovered to my chagrin that most of this information had been readily available all along in subject's self-published autobiography. Although most subjects of backgrounding have not written autobiographies, their books or articles on other topics may tangentially reveal many facts about themselves and their close associates, and provide a window on their values and psychological makeup. In addition, that short story or poem they dashed off for a local newspaper contest may reveal secret desires and unresolved psychic traumas in disguised form.

To squeeze the maximum amount of personal information out of any book written by your subject, you must search systematically from cover to cover. The acknowledgements page may give you the names of relatives, friends, and colleagues of subject; his or her agent and editor; and foundations or government agencies that provided financial aid for his or her research. The page that faces the title page may include a list of subject's previous books. The dedication may provide you with the first name of subject's live-in lover or spouse. Indeed, the dedications of successive books may be the archeological strata of subject's relationships. (A famous science-fiction writer once dedicated a novel about a sexy android to a list of thirty-one women, apparently his most fondly remembered sweethearts. Alas for any future biographer, he did not provide their last names.)

The preface or foreward may include autobiographical remarks (or if written by someone else, appreciative comments on subject's work mixed with a few tidbits about subject's personal background). The bibliography may include several of subject's articles that you hadn't heard of before. The works by others that subject chooses to list in the bibliography or cite in the footnotes and chapter notes may provide a window on his or her unspoken ideological biases.

Often the most important resource is the index, which may include numerous page references to the author and to organizations and individuals linked to the author, including those listed in the acknowledgements. Although the index of an autobiography or memoir by subject will be the most useful, the indexes of books in other categories may lead you to implicitly autobiographical passages both in the text and in the footnotes. This is especially true of works of journalism and contemporary history that include memoir-type material in passing. A good example is *The Rise of the Right*, William A. Rusher's excellent history of American conservatism from the 1950s through the 1980s. Rusher, publisher of *National Review*, was a key player in this history, and the index has hundreds of references to Rusher himself, his magazine, and his closest associates. Thus, although the book appears at first glance to be history rather than autobiography, it actually includes in scattered form a rich autobiographical profile of its author. The presence of such material cannot always be inferred from the title or library classification of a book.

## Reviews and Criticism

To find reviews of your subject's books, use the *Book Review Index (BRI)*. This standard reference work published by Gale includes a bimonthly index and annual cumulations. *BRI*'s *1989 Annual Cumulation* includes about 131,000 review citations for about 74,000 works reviewed in over 500 periodicals. There is a master cumulation for 1965-1984 in ten volumes, including 1.6 million citations. *BRI* from 1969 to the present is available online from DIALOG.

Many scholarly and scientific periodicals that review books are not indexed in *BRI*. Such reviews, however, can be found in the specialized indexes covering these periodicals, which can be searched via DIALOG.

When you go to the cited magazine or newspaper to photocopy the review, look at the letters to the editor section in the following issues: An unfavorable review of subject's book may have elicited an indignant reply either from subject himself or from a friend or admirer; a favorable review may have flushed out subject's detractors and enemies to express even fiercer indignation.

For quotes and summaries of critical opinion about an author—including material from books of criticism as well as periodicals—see the various biographical/critical dictionaries published by Gale. Also see *Book Review Digest (BRD)*, which provides excerpts from and citations to reviews of nearly 6,000 English-language books each year. *BRD*'s retrospective volumes date back to 1907 and can be easily searched via the cumulative author/title index. The contents of *BRD* since 1983 are available online and in CD-ROM format.

Note that if you plan to interview your subject, familiarity with criticial reviews of his or her writing is essential: first, because it shows that you take his or her work seriously; second, because nothing will get a writer (and especially an amateur or part-time writer) talking faster than an opportunity to criticize their critics.

## Finding Subject's Articles and Other Writings

To find newspaper and periodical articles, short stories, plays, and miscellaneous writings by your subject, your first step will be to search the relevant full-text, abstract, and index databases. For publications and issues not searchable by computer, the options include print indexes; clippings files; article, book, and resume bibliographies; and special bibliographical works. In the case of journalists who have written fulltime or as regular freelance contributors for an unindexed newspaper or magazine, just look for their byline in each issue on the microfilm or in the bound volumes. (For details on how to search newspapers and periodicals, see Chapter Six.)

If your subject is a journalist, a database search may turn up hundreds of articles; if a scientific researcher or scholar, scores of articles. This can compensate for the often paltry courthouse paper trail that intellectuals (as opposed to real estate developers and politicians) leave. The articles of a prolific writer comprise a rich chronological record of the topics in which he or she has been interested and the collaborators with whom he or she has worked. In the case of a journalist one gains a fascinating record of the lives he or she has touched and the enemies he or she has made. This provides interesting options, e.g., the Lyndon LaRouche organization, seeking negative information in 1984 about NBC investigative reporter Brian Ross, called up all the mobsters and Teamster hoodlums he had pilloried over the years.

### Critics Of Subject's Articles
While searching for subject's articles, also keep an eye out for replies, which may appear in a letter to the editor format or as a full-blown article. Often such replies will be indexed under subject's name; if not, they can be found by looking through the next few issues of the publication following the appearance of his or her article. The essay or reportage you thought was so brilliant may have been subjected to devastating criticism. If the criticism is less than devastating, at least you've gained the name of someone who dislikes or envies the author.

### Subject's Scientific and Scholarly Articles
In your public library you will find the various H.W. Wilson scientific, scholarly, and professional indexes, including *Applied Science & Technology Index, Art Index, Biological & Agricultural Index, Business Periodicals Index, Education Index, General Science Index, Humanities Index, Index to Legal Periodicals, Library Literature, Religion Indexes,* and *Social Sciences Index.* Most of these works provide coverage back to early in the century, often under a succession of different titles. Several of them lack a separate author index, but database versions enable you to search for an author's name for the covered years (unfortunately none of the databases except the one for *Religion Indexes* begin earlier than 1981). All are available at public libraries on CD-ROM; online access can be gained through Telebase's EasyNet gateway. To search all Wilson databases at once for any author's name, use the Wilson Name Authority File (online only).

Through DIALOG, you can access database indexes with much broader coverage than the Wilson ones (and often going back many years earlier) in every major field of scholarship, science, and technology and in every major profession. Some of the most important are:

- Arts & Humanities Search (indexes 1,300 of the world's arts and humanities journals; 1980 to present)

- ERIC (over 700 education journals; 1966 to present)

- Legal Resource Index (over 750 law journals; 1980 to present)

- Medline (6.2 million records from over 3,000 biomedical journals; 1966 to present)

- Scisearch (over 8 million records from 1974 to present, representing the vast majority of the world's significant scientific and technical literature)

- Social Scisearch (over 2.5 millions records from 1,500 social science journals; 1972 to present).

Several of these and other DIALOG databases cover monographs, multiple-author books, conference papers, conference panel discussions, book reviews, and other special modes of scholarly and scientific communication. Some are also available in print form, e.g, Medline corresponds to three printed indexes: *Index Medicus, Index to Dental Literature,* and *International Nursing Index.*

Note that the *Directory of American Scholars* (over 39,000 entries) may include a list of some of your subject's publications. In addition, you can look in the bibliography of any scholarly or scientific book by subject. Generally the authors of such books are careful to include all of their own articles and research studies relating to the book's topic. (Obviously to gain the biggest list, you would look at subject's books in his or her main specialty, not his forays into other fields.) For instance, research psychologist Theodore X. Barber's *Hypnosis: A Scientific Approach,* contains in the bibliography a list of over sixty articles by Barber written prior to the book's 1976 publication.

A search of subject's articles, monographs, etc., down through the years will provide several types of information to supplement what you find in biographical dictionaries or professional directories. For instance, an article will usually identify the university, think-tank, or government or private-industry laboratory with which subject was affiliated at the time of writing; this may fill in gaps in what you know about his or her job history. By noting the names of subject's co-authors on various articles, you will learn who some of his or her closest colleagues and mentors have been at various career stages. You should also note the names of graduate students or junior scientists whose contributions to a paper were acknowledged in an apparently grudging manner (some of these may feel their work was ripped off or that subject otherwise took advantage of them, and they may be willing to talk about it).

An article or book may also include information about government or foundation grants subject has received. The preface to Dr. Barber's hypnosis book discloses that his research from 1956 to 1976 was supported by grants from the National Institute of Mental Health (MF-6343c, MY-3253, MY-4825, MH-7003, and MH-11521). Once you know the funding source, you can find details on a government grant via Freedom of Information requests; and on foundation grants, via Federal 990 forms or foundation annual reports. Note that grants on federal research projects are preceded by a review of the application by a committee of government scientists. After the research study is completed, another committee of scientists will write a peer review summary report. This report is available from the grantor agency.

Also note that scholarly and scientific indexes usually cover letters to the editors and other communications disagreeing with an article or report. These polemical pieces can sometimes be scathing (as can attacks on subject's works found in the proceedings of scientific conferences). The indignant letter writer or outspoken panelist may become your best source in unmasking the pretensions of a pseudo-scholar or pseudo-scientist.

In searching scholarly and scientific literature (especially if you extend your search beyond the top journals with the highest standards), you will frequently find a large amount of flimflam. Scholars and scientists are under constant pressure to publish or perish (even if they are primarily teachers rather than researchers). The result of this pressure (and of the need to pad curriculum vitae bibliographies to qualify for a better post) is that professors often churn out research articles on trivial topics (or, in the humanities and social sciences, shallow opinion pieces requiring no research). A virtual industry of unrefereed scientific journals and obscure scholarly journals has arisen to abet this practice. In the sciences today, at least 40,000 journals produce a million articles a year, a vast percentage of which is of dubious value.

The various tricks involved include publishing a study in unnecessary installments (with each installment counting as a separate article), publishing two versions of the same findings in different journals, allowing colleagues to piggyback their names onto your articles as co-authors while you also piggyback onto theirs, and of course putting your name first on a study for which your graduate students did all the work. The most extreme padding occurs in the so-called team reports churned out by certain scientific laboratories. These labs will be working on many studies at once, and sometimes a dozen or more lab scientists will be listed among the co-authors of a given study even though the contributions of most of them were minimal. Department heads and research supervisors, in particular, get their names added to scores of articles—a kind of scientific "droit du seigneur."

### Anonymous and Pseudonymous Authors
Many books are published under pseudonyms, but the reason doesn't usually involve any great desire for secrecy. A prolific author may decide that two books under one name in a single year won't sell as well as two books under separate names. A mainstream novelist may decide to write his or her trashy thrillers under a pseudonym in order not to undermine his or her "literary" reputation. Often you can find the truth simply by looking in *Cumulative Book Index* where the author's various pennames are cross-referenced with the real name. Library card catalogs usually have this information, taken from the Library of Congress catalog system. The

Library of Congress card for Wall Street commentator Adam Smith's *Powers of Mind*, for instance, tells us that his real name is George J.W. Goodman.

Things are not so simple if an author is really determined to conceal his or her identity. The Library of Congress Copyright Office's Application Form TX does not require the copyright claimant to reveal the name of the work's author. The claimant of an anonymous or pseudonymous work may in fact be the unadmitted author of the work. But the claimant also may be an employer or other person for whom the work was "made for hire," or someone who simply obtained the contractual right to claim legal title.

Information on copyrights from 1978 to the present is available via computer terminals in the Copyright Office in Washington. For earlier copyrights you must look in the card files. (If you cannot visit the Copyright Office, its staff will research the copyright for a small fee and mail you the information.) If you are searching for the identity of an anonymous or pseudonymous author, examine all the records of the given work: all successive Form TXs, supplementary registrations, and records of transfer of copyright ownership.

If the author is trying to conceal his name (for instance, on a self-published sex manual) and also to retain control of the work, the copyright claimant may be the author's attorney or spouse (especially a wife under her maiden name), or a corporation or unincorporated business set up by the author or an associate. In such cases, the standard backgrounding techniques in this book should enable you to eventually find the author.

Some anonymous or pseudonymous authors will fail to register their self-published book or pamphlet with the Copyright Office at all (although they will probably put a copyright notice on the title page anyway). However, such works will usually have the name of a publisher on the cover or inside, and this publisher—even if it's just a one-book operation—will be a registered business. In addition, if the book is sold mail-order via a P.O. box, you can find out from the Post Office who rented the box (see 12.1).

If you're still stumped, look for clues in the book itself: the printer and/ or typesetter's name, the printers' union bug, the credits for a cover artist or other illustrator (or at least an artist's signature on one of the illustrations), and the photo credits. Any of these clues may lead you to someone who knows the author's identity.

## 11.4 Subject's Garbage

This resource has produced some interesting if smelly results through the years. A private investigator's examination of the trash of Skadden Arps Slate Meagher & Flom, a major Wall Street law firm, led to the arrest and conviction of two inside traders. Freelance journalist A.J. Weberman's probe of the garbage of Bella Abzug during the Vietnam War led to the revelation that Bella and her husband owned stock in two major defense contractors. Jack Anderson's snooping in J. Edgar Hoover's garbage turned up empty booze bottles that called into question Hoover's sanctimonious demand that all FBI agents be strict teetotallers. In the early 1980s I gained much information on Lyndon LaRouche's organization through my garbage rounds (landlords in Manhattan would call me at the newspaper *Our Town* when anything interesting turned up in the garbage of a LaRouche follower living in one of their buildings). Once when a LaRouchian couple moved out of an apartment leaving behind heaps of papers and trash, I spent a happy afternoon crawling around in the dumpster on the street in front of the building collecting bank statements, old phone bills, and supposedly top-secret internal memos of the LaRouche organization.

The *National Enquirer* was criticized for going through Henry Kissinger's garbage. The tabloid's reply: How can Kissinger complain about privacy when he ordered the bugging of his own staff's phones at the National Security Council?

For tips on this unusual technique see A.J. Weberman's *My Life in Garbology*, New York: Stonehill Publishing Company, 1980. You can obtain a copy of this out-of-print classic by writing to Weberman at 6 Bleecker Street, New York, NY 10012.

Weberman cautions against trespassing on private property. Only pick up garbage from public sidewalks. With millions of homeless people sifting through garbage cans all over America today, it's unlikely anyone will challenge you, especially if you dress like a homeless person. Weberman's own specialty was to go to the Manhattan townhouses of the rich and famous where the garbage was either on the sidewalk or in a nearby alley. To make sure he had the right can, he'd sift through a bag in search of a piece of junk mail with subject's name on it.

Of course this can't work if subject lives in a high rise where the people on each floor throw their garbage down a chute to a compactor. To solve this problem, Weberman suggests (tongue in cheek) that you get a friend or acquaintance who lives in the building to let you in some evening. Go to the celebrity's floor and seal down the garbage chute's lid with superglue, so the tenants on that floor will have to leave their garbage bags beside the chute. Then, return in the early morning before the building maintenance staff arrives, and search the bags to determine which is the celebrity's.

## 11.5 License Plate Surveillance

Jeannie arrives at her boyfriend Tom's house earlier than expected. While parking her car down the block, she observes a seductively dressed blonde exit Tom's front door, get in a Toyota and drive away. Jeannie would like to know who that woman is, but she doesn't want to ask Tom and thus tip him off that she knows he has extracurricular visitors.

Russ the anti-Klan activist has stationed himself on a hill above Roy the grand dragon's cow pasture. Through his binoculars he observes several cars and pickup trucks pull up in the yard. Angry-looking men enter the house where they remain for several hours. A Klavern meeting? Naturally, Russ would like to know the identities of the visitors, but he is certainly not going to stroll down the hill and ask them.

If Jeannie and Russ manage to get the license plate numbers of the cars in question, they can get the answers to their questions (or at least clues to the answers) by checking with the state Department of Motor Vehicles (DMV). Motor vehicle registration information is almost always publicly available. If you have the license plate number of a car, you can write to the DMV and get the name and address of the owner for a small fee.

In some states, an answer to a written request may take weeks. You may get a speedier response by using the state's official request form. If you need the information immediately, go through a database information broker. If you need license plate searches on a frequent basis, open an account with the Department of Motor Vehicles so you can get the information over the phone or via online database (one advantage to online access is that you can do cross-searching).

License plate searches involve two basic problems. First, the DMV may send you information on the wrong car. To guard against this, try to note the car's make, year, and color so you can later check this information against what you receive from the DMV.

Second, the driver of the car—the person you are trying to identify—may not be the owner of record. Jeannie, for instance, discovers that the owner of the Toyota is a man; she now must determine if the mysterious woman driver is his wife, girl friend, sister, daughter, or whatever. Russ finds out that the owner of one of the cars at the Klavern meeting is a rental agency. He now must persuade someone at the agency to tell him who rented the car that day.

Depending on the circumstances, other sources of confusion might arise during Russ' ongoing license-plate surveillance of right-wing extremists. One of the cars he observes may turn out to be stolen. Another, parked on a city street, may bear license plates from an abandoned car (this is the driver's little trick to avoid parking tickets). A third, apparently that of an out-of-state visitor to Roy's farm, may actually belong to a local guy who's registered his car out-of-state to get lower insurance rates.

In spite of the above, license plate checks can be quite useful. Essentially there are two types of surveillance involved: you can stake out a residence or business and see who visits, or you can do a drive-through or walk-through, e.g., go to the company parking lot and copy down all the license plate numbers. Both methods may help you find potential sources. A stake-out of visitors at subject's house may give you the name of a maid or relative. A walk-through of the parking lot of subject's small construction company may give you the license plate numbers of all workers in the front office. The uses of such information are only as limited as your imagination and initiative.

When examining parked cars up close, don't miss the information on the windshield parking stickers. Subject's VIP sticker issued by the city may provide your first hint of his or her political and business dealings with the mayor. Some parking stickers will reveal where a person works or goes to school, e.g., the student and faculty stickers issued by universities for on-campus parking. Such stickers are the best way to differentiate the license plates of employees from those of visitors when you take your stroll through the parking lot.

Bumper stickers may also provide useful clues. Let's say that Russ wants to know why a company car from Ajax Plastics was present at the reputed Klan meeting. He goes to the company parking lot and notices pro-David Duke bumper stickers on several cars. But he also notices a car displaying pro-civil rights and pro-choice stickers. The owner of this car will probably be the first person he calls.

### License Plate Codes

License plate letters and numbers—and also the plate colors and embossed captions and symbols—contain coded information. Thus, while observing license plates you can decide on the spot which vehicles to check out further. You can also learn information from the license plates which might not be contained in the abstract of registration. For instance, in several states the license plate letters reveal the county or congressional district in which the car was registered. Depending on the state, a plate may also reveal such information as the use or weight of the vehicle, the first letter of the owner's last name, the owner's occupation or professional status, his or her membership in a particular private organization or Indian tribe, and his or her status as a veteran, government official, handicapped person, or diplomat. Almost always, the plate will tell you if the car belongs to a rental or leasing agency.

New York State license plates are a fascinating example of the above. The dozens of codes include DCH (chiropractor), DPL (diplomatic corps), FC (foreign consular corps), TV (television industry), NYP (press), RX (pharmacist), and PBA (Patrolmen's Benevolent Association). If a New York plate has three numerals followed by a dash and a "Z" plus two other alphabet letters, the car is registered to a rental or leasing company.

License plates may also reveal if a car belongs to the state government and even may identify the particular government department. Who was using the car at a particular time—or who the car is assigned to on a regular basis—should be publicly available information, although you might have to make a formal request under the state's sunshine law.

Federal government license plates are coded for dozens of departments; for instance, "J" means Justice Department and "D" means Defense Department. But many government cars will only have the letter "G" before the numbers, meaning Interagency Motor Pools System.

A detailed description of the coded information for all fifty states, the District of Columbia, and the federal government is contained in Thomson C. Murray's *License Plate Code Book* (see bibliography).

## 11.6 Finding Your Subject in Books and Dissertations

Millions of living Americans are mentioned in passing in books and dissertations. If your subject has a background in politics, big business, national security, or organized crime, check NameBASE (see 6.5), the cumulative database of hundreds of book indexes. If NameBASE is not appropriate for your subject, try the following: First, make a list of all notable persons, organizations, and events with which subject has been associated. Second, collect via catalog databases the titles of books that might have information regarding each name or topic. Third, go to the largest library and/or the most appropriate library subject collection in your city and search the name indexes of all the books on your list that are available; be sure to also examine any lists of interviewees in a book's acknowledgements or appendix to see if subject's (or subject's associates') names crop up. Fourth, see if the library has copies of any other likely books mentioned in the bibliographies of the above books (but which you missed in your database search) and repeat the above process. Fifth, contact the authors of books that mention subject or subject's associates, to see if they have any further information (including interview notes or tapes); also contact the authors of the most promising books not available in the library to see if they mentioned in print or have any information on subject (see 11.3 on how to locate authors). If the author can't be located, call a library that has the book in question and try to talk a staffer into checking the name index for you. In searching dissertations (or master's theses), go to Dissertation Abstracts Online to find the most likely titles, then contact the authors directly with your questions. Note that any dissertation submitted to a local university will be available in the university library. In searching an available dissertation, pay special attention to the lists of unpublished source documents and interviewees in the bibliography and appendix.

# 12·

# Businesses and Nonprofit Organizations

By learning how to background an individual, you have picked up much of what you need to know in backgrounding businesses and nonprofit organizations. Many of the records that apply to individuals, such as UCC filings and tax liens, also apply to corporate entities. An experienced researcher will jump back and forth from corporate to individual files, using the clues found in one to search out information in the other.

## 12.1 Tracking Down Small Businesses and Fly-by-Night Enterprises

As with hard-to-find people, the best place to begin is the telephone directory and city or crisscross directory. If you don't know what city the business is in, consult Dun's Electronic Business Directory, the online directory for over 8.5 million businesses and professionals nationwide. Available on DIALOG, it gives address, phone number, Standard Industrial Classification (SIC) codes and descriptions, and employee size ranges. If you don't have access to a computer, check Dun's Business Identification Service, a microfiche register covering 10.2 million businesses (available in most research libraries). Both these resources are compiled from Dun & Bradstreet credit reports. If a business is listed in either resource, D&B will usually have additional information on it (see D&B credit reports, 12.2).

If you have an address for the business but cannot find a phone number listed in its name, visit the premises or look in the city or crisscross directory to see what other businesses or individuals are listed there. You may discover that your target business is a subsidiary of, or a registered business name for, another business at that location. If the location is a private residence, the business is probably being operated by someone in the household using their private phone. You may also find that the address is that of a corporate registration service (see below) or simply a mail forwarding service, i.e., a "mail drop."

A shortcut for checking mail-drop addresses is the *Directory of Mail Drops in the United States and Canada* (see bibliography), which lists hundreds and describes the services provided by each. Although this booklet by no means includes all of such businesses in North America, it is sold by mail-order book sellers who cater to tax evaders, scam artists, and criminal fugitives. Thus, a business that chooses one of the listed mail drops may deserve close scrutiny.

Some companies offer not only a mail drop but also a telephone answering service and the part-time or occasional use of an office to impress a potential client. (In New York City, such services can provide the "prestige" of a Park Avenue or Madison Avenue address.)

Be aware that many fly-by-night (as well as legitimate) businesses have an RCF (Remote Call Forwarding) number. In selling this service, New York Telephone describes it as a way of making customers think you have a branch office in a particular city when you really don't—the calls are forwarded without the caller's knowledge to an office in another city. A firm can meanwhile list, as the local address for the RCF number, the office of a mail drop which has instructions not to give out the real location. Note that New York Telephone's operators will not tell you if a number is RCF.

### Post Office Box Renters

If a post office box is used for business purposes—for instance, the sale of printed material or the solicitation of funds—the postmaster of the station is required by postal regulations to give you the name and address of the person or corporation renting the box. This is how you can find the identity of, say, an advertiser in your local newspaper's classified section if the ad provides no name. If the ad does include a business name, the postal records may give you a more recent street address than can be found in the business name or corporate registration files at your county courthouse.

### Postal Mailing Permits

Businesses can obtain several types of mailing permits. If the permit is identified by a number on the envelope (as in the franking of metered mail or the use of envelopes with postage-paid imprints), you can obtain the name and address of the permit holder simply by calling the nearest U.S. Post Office mail classification center (there are thirty-seven throughout the United States). If you want a copy of the permit holder's permit application, you must make an FOIA request to the central post office of the city in which the business is located.

The types of permits include metered mail, pre-canceled stamp, imprint permit, first-class presort, and second-class mailing (the latter is used by newspapers and periodicals). Sometimes the different permit applications for a particular business will be filled out by different individuals—this may give you the name of a silent partner whom you otherwise would not have learned about.

## Business Name, Corporation, and Limited Partnership Indexes

Even the shadiest of businesses needs a bank account for depositing its checks. To open an account (and to comply with state registration laws), an entity might file as (a) a partnership or limited partnership, (b) a corporation incorporated within the state, (c) a corporation doing business in the state but incorporated in another state, (d) a not-for-profit corporation, (e) an unincorporated association, (f) an individual, corporation, or partnership operating under an assumed business name (the assumed name is often called the d/b/a, which means "doing business as"). Note that in some states the names of hotels and motels are treated as a separate item and are filed differently from an ordinary d/b/a.

Corporations and partnerships generally file their registration papers with the corporations division of the state's department of state. In most states you can get the files searched over the phone. You can also get a search done by mail or on a walk-in basis. About 30 states sell magnetic tapes of their files, while over a dozen states offer online access. Through LEXIS you can search the state corporation filings (and in most cases the partnership filings) for California, Colorado, Connecticut, Georgia, Illinois, Massachusetts, Michigan, Missouri, Nevada, Pennsylvania, Texas, and Wisconsin. Access in other states is offered by information brokers who are part of the multi-state net of vendors reselling the magnetic tape data.

Business name or d/b/a certificates are usually filed at the county level in each county in which the entity is conducting business. Copies of incorporation or partnership papers may also be sent to these county files from the state office if the corporation or partnership has offices in the county.

In searching county records, the indexes to the business name, partnership, and corporation registrations may be merged into a single alphabetical listing or there may be separate listings. The indexes may be in computer-printout volumes or may be searchable via a computer terminal. Usually the business name file will contain a copy of the business name certificate telling who obtained it and providing an address for service of process. If the business was subsequently incorporated, the file may refer you to an incorporation file number.

The corporation registration file will include the articles of incorporation (and amendments thereto) as well as certificates of incorporation, merger, or discontinuance and change-of-name certificates. If required, the file will reveal the names and addresses of the corporation's principals; if not, the file may only include the name and address of a registering agent authorized to receive service of legal papers (this will be either the incorporator's attorney or a company specializing in corporate registrations).

The limited partnership index accesses files that are often more detailed, including names and addresses of general and limited partners, the percent ownership of each, agreements for how decisions will be made and profits shared, amendments to the rules, and all changes in the identities of limited and general partners and in the classes (e.g., Class A, Class B) of limited partners.

In searching these files, you should have the complete name of the business accurately spelled. The alphabetical listings in a large city will contain many very similar names (e.g., "Masada Associates, Inc.," "Masada Company," "Masada Corporation," "Massada Florists," "Masada Printing Company," "Masseda Enterprises," etc.). Before you begin to search the index books, familiarize yourself with the alphabetizing system (for instance: are acronyms listed at the beginning of the letter listing, or are they scattered throughout in strict alphabetical order?).

Always look in the file for the name and address of the firm's attorney. If you see the name "Julius Blumberg" on the jacket of the incorporation papers, that is not the attorney, but simply the name of a printing company that prints the legal forms most widely used in many states.

Inexperienced researchers often waste time chasing down the names of apparent company officers who really are only employees of a corporate registration firm. One of the best known of these firms is CT Corporation System, which can register a business in all fifty states. Such firms are used by businesses that wish to operate in many or all states without opening an office in each. A corporation incorporates in one state only—either its home state or else a state with minimal regulation, such as Delaware—and then acquires a certificate to do business in other states. CT Corporation System maintains offices in all states and can guarantee that state and local law is being complied with in each case—it can also provide an address within the state for service of process on the corporation.

Sometimes the registration company's name will be on the registration papers; in other instances, individual employees of the registration company will be listed without the registration company's name appearing. In still other cases, the names of one or more of the actual principals will be listed as an incorporator or officer, but with the registration company still being designated as the agent for service of process. The registration company usually will not give you any information about a principal unless authorized to do so; however, they will forward your request to speak with one of the principals.

The corporation file will usually include any name-change certificates. However, you will not necessarily learn from the file whether the corporation previously existed as a d/b/a business or a limited partnership (L.P.). When you request the file, therefore, you should also request the files on previously registered d/b/a's or L.P.'s with similar names. For instance, if the Markco Corporation, Inc. was registered in 1983 and you see that a Markco Company or Markco Associates was registered as a d/b/a the previous year, the latter may be the predecessor entity. If you search the file for this entity you may uncover the name of a previous owner and/or previous attorney. And while you're at it, look for a Markco Foundation or Markco Trust set up for tax purposes.

If you are trying to chase down the various dummy names used in a sensitive real estate deal or a white-collar scam of some kind, the above indexes can be very useful. Often a complicated deal will involve a multiplicity of entities each of which will only be used for a single transaction in a chain of transactions. You can sometimes spot the connection if all of the registrations of these entities were filed during the same narrow time span by the same attorney. The key to this is the file numbers, which generally designate the order in which registrations were filed during each year. If the index books are a cumulative alpha listing without any reverse listing by file number, this order of filing may be difficult to trace. However, you may be able to find the file numbers closest to your target company's by looking in the supplementary monthly indexes. For instance, if the company's file shows that it was incorporated in March 1989, you can ask the file room clerk if the supplementary volume for that month is still available. (Of course, if the index is searchable at a computer terminal the interrelations of various business entities are easy to detect.)

Corporations that fail to file their tax returns within a set period may lose their registration status. A notice to this effect will be placed in the corporate registration file, and the firm may also be listed on a roster of delinquent tax filers.

### Sales Brochures

A business may be extremely secretive around everyone EXCEPT potential customers. In the late 1970s a computer software company controlled by Lyndon LaRouche published a glossy sales brochure that listed dozens of its supposedly satisfied clients. A foray into the waiting area of the computer company's midtown Manhattan office produced a copy of the indiscreet brochure. Likewise, a trip to the office of another business—a Miami export/import firm—produced a copy of a brochure intended for distribution almost exclusively in the Middle East. The brochure included a mail drawer address in the Bahamas that turned out to be the same mail drawer used by a Mafia drug bank.

### Signs On The Door; Lobby and Floor Directories

A friend of mine was trying to trace the business affairs of a Ku Klux Klan-linked toy distributor who for many years had been running several small businesses out of a suite of offices in a Manhattan office building. My friend went to the building and copied from the lobby directory all the business names for that suite. Then he went up to the floor the suite was on, and copied the names from the floor directory. Then he went to the door of the suite and copied the names listed on the frosted glass. The listings were somewhat different in each case, and he ended up with nine names. He then headed for the business names index at the county courthouse.

This was admirable attention to detail: My friend recognized that each of the three sets of signs would have been put up at different times (it was an old building), and he read them off like geological strata.

### Clues In The Office Or Waiting Area

If you visit a company's offices on a pretext, note any business permits or licenses, diplomas, or testimonials framed on the walls—this will give you a fresh paper trial to follow. Note the receptionist or secretary's name-tag or the name plate on his or her desk—this person may become a source later. If there is a calendar on the wall, note the name of the company that gave it as a gift—this may be the business's printer or typesetter or insurance underwriter. If there are magazines in the waiting area, note the name and address on the subscription label—this may give you the home address of a company employee and/or it may be addressed to a company officer whose name you would not otherwise find. You may also discover in the waiting area various company promotional brochures (see above), an annual report, a copy of the company's internal newsletter, and other useful materials. The company's inhouse phone list or directory may be on the receptionist's desk or beside the waiting area phone provided for the convenience of visitors. In some offices, a copy of the latest internal newsletter or phone list will be tacked to a bulletin board above a water fountain or beside the coffee machine or in the snack room; look here also for state and federal Labor Department notices and OSHA occupational injury and illness notices.

### Complaint Agencies

Sleazy, fly-by-night businesses leave behind irate consumers who often will lodge complaints with public and private agencies. Check first with your city and state consumer protection agencies, both of which will usually give you information regarding a company's complaint file over the phone. For further information, call your local Better Business Bureau.

If the business operates under either a city or state license or is subject (as in the case of the professions and some trades) to a licensing board or other oversight agency, see if there have been any complaints or disciplinary actions. Also check if summonses have been issued (e.g., a summons for health violations in the case of a restaurant) and if the business has ever been closed temporarily for violations.

### Special Telephone Numbers

Your regional or local telephone company will offer a variety of audiotex services—everything from "phone sex" and singles talk lines through the number you call to get the latest weather report or a harangue from a skinhead "hate tape." You can obtain a copy of the local audiotex services directory from the phone company for free. For instance, NYNEX has a

62-page directory for the New York City area that lists phone numbers numerically by type of service, and gives the vendor's name and address, a contact name and phone number, and the name of the program. If you want to know who's behind such programs as "New York Hot Date" and "Super Sluts 3," this is where to look first.

Many sleazy businesses (such as those promising miraculous credit cards to people with poor credit ratings) get a temporary "900" number to operate their scams. You can find the name and address of whoever obtained the "900" service by dialing a special directory number provided by the appropriate long distance telephone carrier.

## 12.2 Backgrounding an Established Corporation

For established businesses in America there is a vast wealth of research sources. What follows is a capsule description; for full treatment of the subject, consult Lorna Daniell's *Business Information Sources* and other business reference works in the bibliography; also see the latest DIALOG and LEXIS/NEXIS catalogues. Note that for any corporation of size in our tightly interrelated economy, all good research work includes parallel, indirect, and operative backgrounding (see 1.3, 1.4, and 1.5). At the outset you should define subject company's relationship to the company that controls it and/or the company or companies that are controlled by it or are under common control with it.

### Business Directories and Databases

Sketches of the financial status and types of activities conducted by a business, as well as the names of directors and officers and other information, are available online and in various published directories. Check as many of these sources as possible. If you are using the print directories, do not neglect the weekly, monthly, or quarterly supplements that keep them up to date.

The following list includes databases available through DIALOG (some are available from other vendors as well) and either the print equivalent of the database or the most closely related print directory from the compiling publisher if such exists. Note that with DIALOG and other online vendors, you can do global searches of multiple files grouped in a variety of combinations.

- D&B—Dun's Market Identifiers: Information on 2.3 million businesses with five or more employees and $1 million or more in sales.

- D&B—Dun's Financial Records Plus: Financial statements for over 650,000 businesses going back as much as three years and including balance sheet, income statement, and business ratios for measuring solvency, efficiency, and profitability. Also included are capsule descriptions of the history and operations of another 1.2 million businesses.

- D&B—Million Dollar Directory: Current business information on over 60,000 companies with net worth of $500,000 or more. This information is also available in the *Dun & Bradstreet Million Dollar Directory*.

- PTS F&S Index: Covers company and industry information from 1972 to present (almost 3.5 million records). It corresponds to the print volumes of the *Funk & Scott Index*.

- Thomas Register Online: A directory of over 50,000 classes of products and services with over one million source listings representing 148,000 companies. Also includes over 110,000 trade or brand names. The print version is the 23-volume *Thomas Register of American Manufacturers* (14 volumes of products and services listings, 2 volumes of company profiles, and a 7-volume catalog file).

- Company Intelligence: This database covers financial and marketing information on about 100,000 companies, with up to ten of the most recent news items on each company from more than 3,000 newspapers and periodicals. Included are companies in *Ward's Business Directory* (companies with a gross annual sales volume of over $500,000) and also newsworthy smaller companies.

- Standard & Poor's Register—Corporate and Standard & Poor's Register—Biographical: The first is a directory of over 45,000 public and private companies with sales in most cases of over $1 million; the second includes brief sketches of executives and directors of firms listed in the first (over 68,000 records). The print version is *Standard & Poor's Register of Corporations, Directors and Executives*. Using either version, you can trace the overlapping directorates of various firms.

- Standard & Poor's Corporate Descriptions and Standard & Poor's News: The first file contains in-depth sketches of over 12,000 publicly held corporations; the second, general news and financial information (including annual reports and interim earnings reports) on most of these from 1979 to the present. The print equivalents are, respectively, *Standard & Poor's Corporation Records* and *Standard & Poor's Corporation Records Daily News and Cumulative News*.

- Moody's Corporate News—U.S. and Moody's Corporate Profiles: The first of these two files contains summaries of news on 13,000 publicly held companies from 1983 to the present; the second, detailed information on companies listed on the New York and American Stock Exchanges and about 1,300 other important companies (almost 4,000 records in all). The print equivalents are *Moody's Industrial Manual*,

*Moody's Bank and Finance Manual, Moody's Public Utility Manual,* and *Moody's Transportation Manual.*

- Corporate Affiliations: A corporate family linkage directory for searching out the subsidiaries of a parent company or the parent of a subsidiary. Also contains summary information on each firm (43,000 records). The print version is *Directory of Corporate Affiliations.*

### Company Credit Reports

Dun & Bradstreet's company credit reports on over 9 million business locations are available online to direct subscribers; other D&B credit information is available through CompuServe or NewsNet. TRW business credit reports are available through CompuServe and NewsNet; the latter offers trade payment histories on over 13 million business locations.

### Securities and Exchange Commission Filings

If the business you are investigating is a publicly held corporation registered with the Securities and Exchange Commission, then detailed information is available on the public record. Filings required by the SEC include stock offering prospectuses, proxy statements, the annual 10-K and quarterly 10-Q reports to the SEC, copies of annual reports to shareholders, and many other documents. In these, you will find out who the major stockholders are, who the officers and directors are (with biographical data on each), and who the attorneys and outside accountants are, as well as gaining information on the firm's lines of credit. You will also find audit reports detailing the firm's current financial status and capital structure, brief assessments of all legal suits in which the firm is involved, and information on the firm's subsidiaries, both domestic and international. The 10-Ks and the annual reports to shareholders may include valuable information about the firm's major clients and the government contracts it has received. The proxy statements will reveal the holdings of directors and officers, institutional investors, and beneficial owners of 5 percent or more of the company's stock; and also the details of any in-house loans or other financial transactions between the firm and any of its directors or officers. Always examine Form S-1, the registration statement filed by a company when it first comes under SEC jurisdiction; it contains details about the firm's finances and about the personal and business backgrounds of its officers that may not be repeated in subsequent filings.

By tracing SEC filings by a company over the years you can gather the names of officers, directors, and beneficial owners who are no longer associated with the company—these may be potential sources. You can also find out about persons who have made unsuccessful bids for control and those who have been squeezed out by hostile takovers; see Schedule 13-D (Report of Securities Purchase), Schedule 14-D1 (Tender Offer Statement), and Schedule 14-D9 (Solicitation Recommendation).

The annual *Directory of Companies Required to File Annual Reports with the Securities and Exchange Commission* will tell you if your target company is required to file. SEC filings going back to the beginning of a company's filing history may be available on microfiche at the SEC regional library. The filings are also available at university business school libraries or the business division of any major research library.

For online searches of SEC filings, LEXIS's Federal Securities Library offers the full text of 10-Q and 10-K filings, proxy statements, and annual reports to shareholders since July 1987. The Disclosure Database, available on DIALOG, provides extracts from 10-Ks, 10-Qs, and 20-F financial reports, as well as registration reports of new registrants. The Disclosure/Spectrum Ownership database reveals ownership data about publicly held companies, including institutional, 5 percent beneficial, and corporate insider holdings. Summary information of a company's past SEC filings can be obtained from the Standard & Poor's and Moody's databases described above.

For help interpreting the financial data in SEC filings, see works cited under "Business Research" in the bibliography.

### Annual State Corporate and Securities Filings

All states require domestic corporations (i.e., those incorporated in the particular state) to file an annual report with the Secretary of State. These reports are more or less revealing, depending on the state's laws and regulations. Generally if you study a corporation's annual reports going back to the first year of it's incorporation, you can get some idea of its history: changes in address, changes in attorneys, changes in officers and directors. If the corporation has changed its name, the name-change papers will usually be in the file. If the annual reports filed by the corporation under its old name prior to the change are not in the file, be sure to ask for them.

The paper trail of annual reports will often turn up lively informants. A former officer may have quarreled with the principals and thus be willing to tell all. The office building manager at the corporation's former address may have information on its financial problems. In addition, the names of the firm's past and present attorneys may be a tip-off to secret corporate or organized crime links.

Note that corporations selling securities within a given state must file a variety of disclosure statements with the state securities regulator. These disclosures can be almost as revealing as SEC filings, and should always be checked when backgrounding a local corporation.

### Newspaper and Periodical Business News

Start with NEXIS. It offers full-text searches for keywords (the names of your target business, its subsidiaries or parent, and its officers and directors) in hundreds of city, state, regional, and national business newspapers and periodicals, newswires and newsletters, and the business and general news sections of hundreds of metropolitan dailies and popular magazines. Is your target firm in the Seattle area? The Business Files of NEXIS include the *Puget Sound Business Journal* (from 1985), the *Pacific Northwest Executive* (from October 1986), and *Seattle Business* (from February 1986), while the Newspaper Files include the *Seattle Times* with its daily business section. Does the company manufacture aircraft components? You might find vital information in the NEXIS Newsletters Files, which include *Aerospace Daily* and *Aviation Daily* from January 1989; or in the Magazine Files, which include *Aviation Week & Space Technology* since January 1975 and *Aerospace America* from January 1984. Is the firm designing high-tech components for Navy jets? The Trade/Technology Files include *Defense Electronics* since January 1982 and *Electronic Business* from January 1988.

If you don't have access to NEXIS, try DIALOG, which offers Trade & Industry Index and Trade & Industry ASAP. The former provides complete indexing and selective abstracting of over 300 publications and selective indexing and abstracting of another 1,200 publications (1981 to present). It also includes the complete text of more than 85 of these journals. Its companion database, Trade & Industry ASAP, provides the complete text of selected articles (from 1983 on) for more than 200 of the covered publications.

Rather than searching databases, you can check the published indexes in your local library. Most important are the *Business Periodicals Index*, which covers 344 business magazines and has retrospective volumes back to 1959, the Funk & Scott domestic and international indexes, the *Wall Street Journal Index*, and the *New York Times Index*. Although all of these are available online, a relatively efficient search can be done in the print volumes because the latter include indexing by company name. Even relatively obscure local businesses will be found if they were ever the subject of a news item relating to development of a new product, Chapter 11 reorganization, a liability suit, etc.

There are thousands of trade journals covering every type of business, most of which are not indexed in the *Business Periodicals Index* or the Funk & Scott indexes. Some you will find on LEXIS, but often not going back far enough. You can find the names of the more obscure publications in *Standard Periodical Directory*. If a periodical you want is not in a local business library, check the local branch of the trade association in question; they may have it on file. Or you can contact the editors of the publication directly—they may agree to search their own files and send you clippings in return for access to your findings. They might also provide you with a bit of unpublished gossip about your target company.

### Brokerage House Reports
Such reports are based on interviews with corporate officials and the individual judgment of the analyst preparing the report. They contain facts and analysis focusing on a firm's clients, management difficulties, govern-

ment contracts, and future plans. These reports are available in business libraries or online via the LEXIS Company Library, which includes reports from over 110 international, national, and regional brokerage firms. For further information on a given company, call the analyst who wrote the report.

### Lexis Law Libraries
Specialized LEXIS law libraries enable you to search online a vast wealth of court decisions, administrative rulings, regulatory commission decisions and government filings (as well as newsletters and bulletins reporting thereon) in over thirty subject areas. You can either do a global search of an entire library or look through various special file combinations. Some of the libraries are specific to a particular group of industries, e.g., the Federal Communications Library and the Federal Transportation Library. Others deal with specific problems that face industries across the board; for instance, the Federal Tax Library, the State Tax Library (files for each state), the Federal Labor Library, and the Federal Bankruptcy Library. In studying any company that does business with the government, you will want to consult the Federal Public Contracts Library. Note that the LEXIS law libraries often range far beyond the law itself. For instance, the Federal Patent Library offers broad access to technological data on various industries, e.g., the file CMPTRS (Computers) which includes scores of computer journals dating back as far as 1982.

### Government Regulation: Audits, Inspection Reports, Enforcement Proceedings
As you gather government data on a corporation step-by-step, keep asking yourself what other government departments or regulatory agencies might require it to file or might be investigating it. Information may be available from dozens of government sources. Is your target business a limited partnership constructing federally subsidized housing? There will be reports and audits available from HUD, which can be obtained either via a Freedom of Information request or via your local congressman. Is a computer hardware firm doing extensive business with civilian federal agencies? GAO reports may be available. Is another computer firm doing business with the Pentagon? The DOD's press office can provide you with a computer printout—unless the data is classified—of all the firm's current and past contracts, broken down as to general purpose, location, and amount of money involved. You can then request the DOD's reports on how well the firm has fulfilled its contracts.

### Environmental, Consumer, and Social Policy
Federal law requires every manufacturer with ten or more employees to report to the Environmental Protection Agency and state agencies the total amount of each of about 330 toxic chemicals it has released into the environment. The data is listed in the EPA's annually published Toxics

Release Inventory, available on microfiche at federal depository libraries (if you're a reporter on deadline, call the EPA at (202) 260-1531 for information on a specific company). Also see your target company's SEC filings: In 1989, the SEC ordered companies to begin disclosing on their 10-K's any potential liabilities they might face under federal environmental cleanup laws. The issue of such liability has helped to spark a number of national and regional commercial databases specializing in the environmental practices of companies as reflected not just in their 10-K's but also in the huge number of documents filed with state and federal environmental agencies each year. Note that the LEXIS Environmental Law Library includes not only environment-related court decisions and administrative rulings, but also the full text of publications that carefully monitor the public record, such as *Environment Reporter* and *Asbestos Abatement Reporter*.

Both federal and state levels of government are a gold mine of information about consumer complaints and product safety. An excellent guide to these sources is Lesko's *Info-Power*, which describes, for instance, the various databases maintained by the Consumer Product Safety Commission. Many state consumer protection offices will give you information about a company's complaint record over the phone (a list of the offices for each state and their disclosure policies is in Lesko's *Info-Power*).

Corporate misdeeds are monitored by numerous public-interest groups. The Manhattan-based Council on Economic Priorities collects data regarding such issues as corporate support for charities, job equality for women, and animal testing. Corporate Democracy, Inc., also based in New York, monitors corporate accountability to stockholders. The various state Public Interest Research Groups and Ralph Nader's Washington, D.C.-based Center for Study of Responsive Law are strong on consumer issues. Scores of other nonprofits that monitor the corporate world can be found in the *Encyclopedia of Associations*—and don't forget state and national trade and industry associations, which sometimes keep files on the most egregious offenders in their ranks.

### Bankruptcy Files

Whether a business is large or small, it may at some point have applied for reorganization under Chapter 11, either under its present name or a previous one. The bankruptcy files will contain a vast wealth of information about customers and clients, vendors and suppliers, claims by or against the firm, pending litigation in any jurisdiction, mismanagement, and possible fraudulent practices. The Chapter 11 status may continue for several years, during which period the firm must file extremely detailed reports on its business affairs. For a local firm, check the bankruptcy index at your federal district court. For firms in other jurisdictions, check LEXIS and/or contact the appropriate federal court clerk's office.

## Licenses and Permits

Most businesses operating in a given city will be subject to city, state, or federal licensing and permit provisions. Consult your local government handbook and/or call the city and state licensing agencies to find out which requirements apply to your target business.

## Miscellaneous Public Records

Check the state and federal court indexes, the federal tax lien files, and the county judgment docket. Also check the statewide UCC filings to get a better picture of a company's liabilities.

## Official Company Sources

Even relatively small companies publish sales brochures, annual reports, etc., and issue press releases from time to time. Such literature can usually be obtained easily from the company's public relations office. Press releases carried by Business Wire since 1983 can be accessed via NEXIS; since 1986, via DIALOG. If the information in a press release is at all significant, it will usually have been included in a news article for some business or trade publication accessible online.

The slickly printed annual report that a firm circulates to stockholders and the general public should not be confused with the 10-K annual report sent to the SEC by publicly held corporations. Although the printed report may contain some hype, it will also include a financial statement, a list of directors and officers, information about major new contracts, etc.

Another good information source is the firm's house organ. See the *Internal Publications Directory*. If your target firm is not listed there, call the firm's public relations director. In general, PR directors will cooperate with business students or freelance business writers. You might ask for an interview with the PR director and, in that context, ask for access to company publications of various types. Another possible route would be through the firm's outside PR consultants; under some circumstances these might be more cooperative than the in-house PR director.

## Former Officers, Directors, Or Managers

Former officers or directors who resigned or were sacked may be willing to discuss the firm's past. To get their names, look in old SEC filings or in old editions of *Standard & Poor's Register of Corporations, Directors and Executives*. In the case of a smaller company, check its annual state filings or the certificates of amendment or merger in its corporate registration file. If a name has disappeared from the list of directors or top executives, it is almost certain the person is no longer affiliated with the firm. Of course this will only help you find the former top managers. For lower-level former managers, try the following:

- Look in back editions of the local city directory for people once listed as employees who are no longer listed; if the old directory does not tell what their job title was, you can estimate from the income level of their neighborhood whether or not they were likely to be on the management level.

▪ Get an old issue of the firm's in-house phone directory and call the extensions of people with management titles to find out which of them no longer works there.

▪ Do an online search of the Federal Election Commission's Contributor Search System to get the names of all employees of the firm reported as having given substantial donations in past years to political candidates or PACs (especially the corporation's own PAC). Most of these contributors will probably have been management-level employees at the time. Then search for more recent contributions from each of these contributors—this will provide you with the names of those who, at the time of the more recent contributions, no longer listed themselves as employees of your target corporation. (Note that the FEC databases cover elections back to 1977-78 and usually give a contributor's place of employment at the time of a given contribution.)

To find the current whereabouts of former executives, directors, and managers look in the current edition of *Standard & Poor's Register*, the local city directory, or the relevant professional or trade directories. Generally, former corporate officials will be easy to find.

### Labor Union and Shop Floor Sources
If the employees of a firm are unionized, then shop stewards, the union business agent, or officers of the local should be contacted. In their preparations for contract negotiations, they have to keep well informed about the profit margins of the business—to know what the traffic will bear. If the union is locked in a bitter dispute with management, it may be willing to help you in hopes that you will turn up something useful for its own purposes.

Even if the firm is not unionized, it is important to contact workers on the shop level. They will be familiar with who the firm's suppliers and customers are, whether orders are up or down in recent months, whether management is covering up shoddy production standards or violations of environmental laws, etc.

Contacting workers if the firm is nonunion or if the union leadership is uncooperative can be tricky. Obviously you can't just walk up to employees on the factory floor and start interviewing them. One way to find the names and home addresses of workers is through a city directory that lists each householder's occupation and place of work. Another method is to copy down license plate numbers of cars in the company parking lot, then get the owners' names and addresses from the Department of Motor Vehicles.

Of course with the above approach, you might inadvertently hit on a

supervisor who'll immediately inform his superiors. Dashiell Hammett's
Continental Op in the classic detective novel *Red Harvest* had a way
around this: When he arrived in town the first person he looked up was
the local leader of the IWW (Industrial Workers of the World, or "Wobblies"). Today there are almost no Wobblies left, but there will often be a
scattering of workers who have engaged in unsuccessful unionization
drives or, if the plant is unionized, who belong to rank-and-file union
caucuses more militant than the union leadership. There will also sometimes be one or two radicals affiliated or formerly affiliated with left-wing
parties. (These radicals will almost always be involved in the rank-and-
file caucus, so if you find one you find the other.)

If you're doing an investigative journalism piece on a particular corporation in, say, a medium-sized city in the Midwest, I would suggest contacting the best known longtime radicals in that city as well as pro-labor
journalists for the local alternative weekly, local ministers with pro-labor
sympathies, or local environmentalist groups. At least one of these sources
might put you in touch with activists inside the plant. But in making these
inquiries, you, as an outsider, should not press for the names of plant
workers; instead ask your local contacts to have the workers get in touch
with you.

Note that in several industries there are national rank-and-file groups
at odds with corrupt union leaders. To find out the situation in a union
shop it may be best to go through such a group. For instance, if I were
backgrounding a Midwest trucking firm, the first place I would go would
be Teamsters for a Democratic Union, which is extremely well-informed
about the business woes of various trucking companies.

Also look for former shop-level employees with a reason to be angry
at management: those who were fired on trumped-up charges after attempting to unionize, those who lost their jobs during the latest cutbacks,
those who were forced into early retirement and now find their pensions
almost worthless in the wake of a corporate takeover. Such people may
be more willing to talk than workers still clinging to their jobs. To find
them (if union leaders won't help and you don't have any local sources),
you might look in back issues of the union local's newspaper for news
about firings. Or you could look in back editions of the local city directory
and compare it with the current edition to see who's no longer at the
plant. Of course, many sacked workers will have left town to find new
jobs.

Note that some unions have auxiliary organizations for retired members. A researcher gathering information on past labor-management disputes at a certain union shop might gain the enthusiastic cooperation of
local members of the retiree's association.

Another approach is to zero in on workers whose disputes with management are a matter of public record. You should check the local courts for
lawsuits by employees or former employees against the company. You
should also do a full-text search for the company's name in LEXIS' Federal

Labor Library, which includes National Labor Relations Board decisions from January 1972, Equal Employment Opportunity Commission decisions from January 1970, and Occupational Safety and Health Review Commission decisions from November 1971.

### Competitors, Suppliers, and Industrial Customers

To find all the companies providing a particular product or service, look in the fourteen-volume Products and Services section of the *Thomas Register of Manufacturers*. There you will find the companies offering each of 50,000 products and services, listed by state and city within each product category. This will give you an idea of who your target company's competitors are. To figure out who its suppliers are (or the companies that it is supplying), you will have to learn something about the particular processes of the industry and then study carefully the relevant product or service listings and consult the seven-volume catalog section. In some cases, the supplier nearest your target company may be the most likely one if transportation cost is a significant factor. For some services, a nearby location will be likely because of the need for personal visits. (In figuring out who the smaller suppliers or customers might be, you may have to check state manufacturing directories and/or the yellow pages since the *Thomas Register* only includes about 145,000 firms.)

As you gather lists of probable suppliers, competitors, or customers, look in LEXIS and also in your local court indexes for any litigation between any of these firms and your target company. Also look in the local corporate judgment books for any judgments obtained by your target company against a customer or supplier (or vice-versa).

## 12.3 Investigating a Nonprofit Entity

Nonprofit corporations often own real estate, are plaintiffs or defendants in lawsuits, become the target of civil judgments, and engage in disputes with their unionized employees. Thus, in researching a nonprofit corporation, you should first read over the preceding section on backgrounding businesses.

### Not-For-Profit Corporate Registration Papers

Nonprofit entities must file their certificates of incorporation, amendment, change of name, or dissolution with the state's department of state. Copies may also be found at the county clerk's office.

### State Charity Filings

Nonprofit organizations must file annual reports with the state division of charities under certain circumstances (in New York, if they solicit more than $10,000 in the given year).

### State Attorney's Office

In most states, a charities unit within the State Attorney's office is responsible for investigating violations of the laws governing nonprofit corporations. If complaints have been lodged against a nonprofit, this bureau may have an extensive file on it.

### Federal 990 Forms

Tax-exempt nonprofit organizations must file an annual 990 Form with the IRS. To see if your target organization qualifies as tax exempt, check the IRS's two-volume *Cumulative List of Organizations.* All 990 filers must make their three most recent 990s available to the general public at their principal office during regular business hours. (The rules are somewhat different for the 990-PF filed by private foundations—see below.) Disclosure applies to all parts of the form except contributor lists.

An organization's earlier 990s can be obtained from the IRS via your district office, as can the latest 990s if you can't visit the organization's headquarters and it refuses to honor telephone or mail requests. However, the IRS can take as long as six months to send you the forms and often claims it simply can't find them. Fortunately, copies are often filed with the state bureau of charities along with the state-required annual report. In requests to various states I have usually received the 990s within a week or so. When you try the state division of charities, also try the State Attorney's office; if it has a complaint file on your target organization, this file may also include copies of the 990s.

Note that if you request a nonprofit's 990s from the IRS always make your request under IRS Code section 6104 rather than the Freedom of Information Act, and be sure to include the organization's Employer Identification Number.

### Records On Private Foundations

A tax-exempt private foundation has a somewhat different legal status from a public charity or a nonprofit institution such as a church or university. To qualify as a tax-exempt private foundation, an entity must receive contributions from only a very limited number of contributors (a single family, for instance) and must make grants only to nonprofit organizations, not to individuals. If it chooses, it may operate its own charitable or public-service programs. There are currently over 30,000 private foundations in the United States.

All private foundations must file Form 990-PF with the IRS annually. This form provides an extremely detailed picture of the finances of a foundation and how it spent its income during the previous year. The form also includes a list of officers and directors, the salaries of top officers, and information on any foundation political activities or any changes in control of the foundation.

The IRS requires private foundations to make their most recent 990-PFs available to the public for 180 days each year; during that period they must provide free copies upon request. If you need to examine previous forms or the current form after the 180-day period has passed, many foundations will gladly send you a copy. If not, you should check with the regional office of the Foundation Center, a private organization that collects past and current 990-PF forms as well as printed annual reports of many private foundations. If the Center doesn't have the reports you need, you can order them from the state division of charities or the IRS.

### Contributor Lists

Although not required by law to reveal their contributor lists, many nonprofits will publish donors' names to enhance giving as well as to satisfy the vanity of the donors. These lists are found in everything from promotional pamphlets to annual reports and even in program publications of the local symphony orchestra. The contributor lists of cultural institutions as well as of universities or hospitals generally include many corporate donors.

### Directories Of Foundations and Charities

Brief descriptions of most U.S. private foundations are given in the Foundation Center's *National Data Book*. The Center also produces the *Source Book* (detailed profiles of the biggest 1,000 foundations) and the *Foundation Directory* (sketches of over 6,500 foundations with $1 million in assets or a minimum of $100,000 in annual grants). The most useful Center publication is the annual *Foundation Grants Index*, which tells who got what, with indexing by key word and name of recipient.

Other directories in this field are published by Gale Research Inc. and its subsidiary The Taft Group. These include *Charitable Organizations of the United States*, with sketches of almost 1,000 fundraising charities and a critical evaluation of each charity's expenses for administration and fundraising as opposed to program; *Taft Foundation Reporter*, with information on the history and priorities of the 500 biggest private foundations (and biographical details on the donors of each); and *America's New Foundations*, with profiles of over 800 foundations set up since 1984 and the names and addresses of 1,700 more established since 1987.

### Foundation Databases

Via DIALOG you can access the *Foundation Directory* database, which describes over 25,000 grant makers. DIALOG also offers a cumulative version of the *Foundation Grants Index* with records from 1973 to the present (this file currently includes over 400,000 records with 20,000 new ones added each year, but does not include grants to individuals or grants totaling less than $5,000).

### Foundation Center Libraries

The Foundation Center operates four regional libraries—in New York City, Washington, D.C., San Francisco, and Cleveland—while affiliates of the Center maintain smaller collections at about 175 public libraries and other locations around the country, focusing on the records of foundations in the given locality. Although Center libraries are open to the public, the library staff will not answer complicated telephone or mail inquiries unless you are a representative of a member organization.

The Center does not maintain systematic records on public charities, tax-exempt fundraising organizations, or tax-exempt institutions such as churches and private universities.

### *Watchdog Organizations*

The National Charities Information Bureau monitors over 400 national charities. It has information on conflicts of interest, fundraising tactics, budgets, and boards of directors. NCIB is located at 19 Union Square, New York, NY 10003.

The Council of Better Business Bureaus' Philanthropic Advisory Service monitors nonprofits that raise funds from the public. It has records on about 7,000 entities. The address is 4200 Wilson Boulevard, Arlington, VA 22203.

# 13.

## Indirect Backgrounding

### 13.1 Finding the "Experts"

In looking into a topic on any but the most superficial level, you will develop questions that can't be answered easily from books. These will often be very detailed and subtle questions that require a chat with an expert. Indeed, you may need the expert to steer you to the books. An expert can rattle off the names of articles and books on the phone that you might only find by luck via library research.

To locate experts and eventually find *the* expert with the most relevant information, you can draw on a wide variety of sources:

- Public-interest advocacy groups, trade and professional associations, and other nonprofit organizations are almost always willing to help you. To find the most relevant groups, look in the subject and location guides of the *Encyclopedia of Associations* and the *National Trade and Professional Associations of the United States*. Contact the research director or newsletter editor of the group. He or she may have the information you need, or may steer you to another staff member or an outside expert.

- The *Directory of Experts, Authorities & Spokespersons* is the annual volume used by radio and TV talk-show hosts to find people to appear on their shows. It contains a number of offbeat specialties, such as UFO research, not easily found in other directories. As appearing on talk shows doesn't pay any money, the presence of a person's name on this list suggests that he or she is not stingy with his or her time, but is an enthusiast eager to share knowledge.

- Many veteran journalists swear by the Heritage Foundation's *Annual Guide to Public Policy Experts*. For some reason, the experts listed herein are unusually generous with their time in telephone interviews, and it has been suggested that the title be changed to the *Guide to Experts Who Love to Talk*.

- *Newsletters in Print* provides detailed entries, arranged by subject, on over 10,000 specialty newsletters. Although the editors of nonprofit newsletters are usually helpful, some editors of commercially produced newsletters (which sell for high prices to very small subscriber lists) are reluctant to share information. Nevertheless, the high quality of their information and files makes them worth a try. If a newsletter editor thinks that you will quote him or her and mention the newsletter in your article, you will get better cooperation.

- The Library of Congress has a computerized directory of almost 14,000 organizations and individuals willing to provide free and fee-based information in a variety of fields, especially in science, technology, and the social sciences. Filed by subject and names, this database gives the organization or individual's specific areas of expertise and also describes special collections, databases, publications, and special services. Call the LC National Reference Service at (202) 707-5522.

- The *Directory of American Scholars* and *American Men and Women of Science* provide listings by academic and scientific specialty and brief educational and career data on each scholar or scientist. The latter is available online from DIALOG.

- The *Research Centers Directory* and its supplement, *New Research Centers*, can help you track down many of the best university-affiliated research experts. At most research centers, the public affairs office will refer you to the appropriate staff member.

- Faculty directories and course guides from local universities will help you to find experts in every field in which courses are being offered. This will include some very practical fields such as real estate finance and hotel administration, as well as the liberal arts. To find out what the specialties of each department member are (if the course guide doesn't tell you), check the *Faculty Directory of Higher Education*, which will list the courses each professor teaches.

  Although local professors will rarely be the ranking experts in their specialty, they will be more likely to spare time for you than will a national academic celebrity. In addition, you can talk face to face with them, and perhaps get access to their personal research files, without having to invest in an airplane ticket. Be aware that local professors often toil away in a narrow but important field. They are often delighted to be contacted by a journalist.

- The catalogs of local alternative education centers and college continuing education programs list teachers with very unusual specialties. Although you are unlikely to need an expert on Hot-Air Ballooning or How to Flirt in Art Museums, such centers also offer courses in very hard-nosed practical subjects. The teachers are often writers, part-time consultants, or editors of specialty newsletters. Because they are into self-promotion, they will often be happy to answer questions from journalists.

- Textbooks and works of scholarship that relate to your field of interest will guide you to a wide range of experts. Look in the bibliography, and also note the authorities cited in the text and in the footnotes or chapter notes. If you find the right book, this is often the quickest way to track down the very best experts.

- Dissertation indexes (see 11.1) can help you narrow your search for an academic expert to the most minute subspecialties. Once you find a dissertation that directly deals with your topic, its author, if located through the *Faculty White Pages*, will probably be so flattered that someone other than his dissertation advisor actually read his turgid tome, that he will cooperate and even give you the names of experts whose knowledge is more up-to-date than his own.

- Investigative Reporters and Editors (IRE) maintains a file of the subject specialties of its 3,000 member-journalists. For referrals, call (314) 882-2042.

   Also check IRE's *Investigative Journalist's Morgue*, an index to thousands of stories and series from the IRE files. This subject index does not give the names of the reporters, but if you find an article or series that fits with what you are looking into, IRE will provide you with the reporter's name and telephone number and a copy of the article.

- In many communities, at least one private citizen collects and files away all the dirt and scandal involving local politicians, businesspeople, etc. Some of these amateur muckrackers are like pack rats, collecting everything they can get on everybody around them. Others are "selective" muckrakers concerned with a particular area. For instance, a corrupt trade union local may have at least one rank-and-filer who occupies his or her spare time with collecting evidence of the union leadership's misdeeds.

   Unlike newspaper staffers—who must flit from assignment to assigment according to the needs of the newspaper—amateur muckrakers have the leisure to concentrate on their pet target(s) year after year, gathering every scrap of documentation they can find. Frequently their files will overflow the basement or attic. Almost invariably, they are eager to cooperate with anyone who shares their interests.

   Let's say you need an expert on Sun Myung Moon's Unification Church. National newspapers will have one or more reporters who cover religious cults, although not too closely. Get their names from IRE, then call them and ask for a referral to someone with special knowledge of the Moonies. They may give you the name of their favorite amateur muckraker as a favor to the muckraker, not as a favor to you. After all, by providing the muckraker with a new potential collaborator (you), they are placing the muckraker in their debt.

Once you find an amateur muckraker, he or she may refer you to other muckrakers, including those who like to keep their identities secret.

▪ Federal government bureaucrats are often remarkably conscientious in helping journalists, corporate researchers, high school science project students, and any other member of the public in need of information. Example: I was backgrounding a businessman who was involved in offshore banking schemes in several small Third World countries. Looking in various directories, I obtained the names and telephone numbers of the State Department and Commerce Department country officers for these nations as well as other relevant officials. None had ever heard of Businessman X, nor would it have been appropriate for them to discuss him if they had. However, they did share with me their intimate knowledge of the countries in question, citing various government and private reports, and they sent me clippings from newspapers in these countries. In addition, they referred me to experts in other government agencies and the private sector. One official referred me to a white-collar criminal in a federal penitentiary whom he described as the best expert on the subject into which I was inquiring. A second official referred me to an investment promoter long active in Country X who was willing to make some discreet inquiries for me.

▪ Congressional committee and subcommittee research staffers are among the most valuable contacts in government for an investigative reporter. The committees for which they work frequently investigate areas such as white-collar crime, government waste and inefficiency, labor racketeering, offshore banking, and corporate pollution of the environment. Quite often, these investigations will shed light on the activities of an individual, corporation, or nonprofit organization in which you are interested.

To find the right committee for your purposes, look in the *Congressional Directory*, and also consult the *Congressional Information Service Index and Abstract*, which is a guide to congressional hearings and reports since 1970. You may find that a committee investigated and held hearings on your topic of interest several years ago, but the committee staffers who did the research have moved to new jobs. Track them down!

▪ Your local congressman may not be an expert on much of anything, but he has a personal legislative staff at his disposal as well as the research staffs of the House committees to which he belongs. In addition, he has the entire resources of the Library of Congress to draw upon through the Congressional Research Service, and he can make the congressional liaison staffs of every department and agency of the federal government jump to his tune. You are his constituent and if you are also an investigative journalist he will have a very strong incentive to keep you happy. Put him to work for you!

- Look in your state and local government handbooks to figure out who's most likely to be useful, both in the various departments and agencies and in the legislative bodies. Remember that state legislative committees, like congressional committees, hold hearings, publish reports and employ staff researchers.

- Tap the energies of retired experts. Throughout America there are millions of retired civil servants, college professors, scientists, and corporate executives with a vast wealth of knowledge and experience in every conceivable field. Many of these retirees are sick of playing golf and would be delighted to cooperate on a volunteer basis with an investigative journalism project that impinges on their specialty. Note that retired experts will often talk more frankly about the Way Things Really Work than might an expert who still has to worry about his or her job. (Thus, if you are looking into industrial pollution, seek out retired chemical engineers who may have spent much of their working lives reluctantly doing the bidding of the polluters.)

## 13.2 Finding the Books You Need

If there is a thorough, up-to-date book on the topic you are researching, you should spare no effort to find it—no matter how obscure the publisher. This lesson was brought home to me in 1979 while researching Lyndon LaRouche's ties to the Teamsters Union. I had been making phone calls for weeks, with only modest results. Then, I happened to mention to a Teamster dissident the name of an obscure (I thought) midwest Teamster official who had been cooperating with LaRouche. "Oh, that s.o.b.," said the dissident, "you can read all about him in *The Hoffa Wars*." He then told me about a book that, if I had begun my research at the public library, I would have found long before. I rushed out and bought the book (Dan Moldea's classic account of Teamster strife) and looked in the index. There, I found dozens of references to the thuggery of this "obscure" Teamster official and his underlings.

In searching for the right book, your first stop will be the catalog of your local research library. But never expect too much. No library, except the Library of Congress, has more than a fraction of the books published in the United States in recent years. In addition, the subject listings in library catalogs will only guide you to the right book if your topic happens to be the book's main topic. In many cases, the best books for your purpose will only deal with your topic peripherally.

Your second stop is the library stacks, if you can gain access. I have often found relevant books—books I probably would not have found through the catalog—by browsing on either side of where the catalog had informed me that a relevant book (usually missing!) was located. Of course this rule of thumb is no substitute for the Dewey Decimal System.

Finally, go to the online catalogs: LC MARC, REMARC, and either RLIN or OCLC. (See Chapter Eleven.)

## 13.3 Finding the Right Library

Don't judge a library only by the size of its total collection. Instead, look at the strength of its collection in the field in which you are interested. If I were researching a left-wing extremist group, for instance, I would not go to the giant New York Public Library; I would go to the Tamiment Institute Library at New York University. Tamiment is tiny compared to the NYPL, but radical labor history happens to be one of its specialties.

To find the strongest collection for your purposes both locally and nationally, look in *Subject Collections*, the *Directory of Special Libraries and Information Centers*, and *Directory of Archives and Manuscript Repositories in the United States*. If the collection you need turns out to be in another part of the country, call to see if the staff will do a search for you. If the library is well-staffed and if its materials are well-indexed—and if you can interest the librarian in your research task—you may have no problem. The staff will find what you need, photocopy it, and send it to you—or lend it to your public library or a local member library of the Special Libraries Association through the relevant interlibrary loan agreements.

The Library of Congress is by far the largest library in the United States. As well as having scholarly books and the most complete periodicals collections in the country, it has all kinds of miscellany, such as pamphlets and brochures, current and back telephone books, and city directories for localities throughout the country. The LC's National Reference Service will help you over the phone with relatively simple questions. If you need complicated research done at the LC, its Reading Room maintains a list of private researchers who will help you for a reasonable hourly fee. Washington journalists on deadline frequently call the LC's public affairs office and set them to work on research tasks.

For your everyday needs, investigate the various libraries in your locality. As noted above, access to the stacks is extremely important. Another important factor is whether or not the photocopying machines work properly and whether or not you have to stand in line to use them. A third factor for any investigator is whether a library has basic tools of his or her trade, including city or crisscross directories, *BGMI*, the very latest editions of other important reference directories (and ease of access to the back editions), and a wide array of CD-ROM full-text-search products. If your local public library can't meet all these needs, you might consider making an arrangement with a university library (these are easier to gain access to than one would think).

# 13.4 The World of Research Cabinets

Nonprofit organizations often maintain clippings files on subjects of interest to their members or sponsors. These files may range from a single filing cabinet to vast clippings libraries (such as the one maintained by the Anti-Defamation League in New York), which rival those of the great daily newspapers. In some cases, these files turn out to be a researcher's dream. They can greatly shorten the time it would otherwise take you to collect articles on your subject from indexed periodicals. In addition, these files will include articles from obscure unindexed periodicals that you probably would never come across on your own.

Often these clippings files have been developed as an adjunct of the organization's library, which may be listed in the *Directory of Special Libraries and Information Centers*. But many small organizational libraries are not listed in the directory, and even when they are, the directory may not specify the existence of clippings files.

The easiest way to find such files is to ask the experts in the field you are researching. Failing this, check the *Encyclopedia of Associations* and also *National Trade and Professional Associations of the United States*. Many organizations listed in these directories regard the dissemination of information to the general public as one of their major functions. Often, they will search their files for you and send you photocopies. Or if you visit their headquarters, they will help you find what you need and provide photocopying on the spot.

If an organization has extensive files relating to your topic and is within convenient distance, you should visit to search the files yourself. Once you meet the staff in person, and explain your purpose in greater detail, they may let you look through files not ordinarily available to the public.

**Example:** I was researching LaRouche's ties to the South African government. I went to an anti-apartheid research organization in New York, which was happy to help. Within minutes I was looking through thick folders of clippings from South African daily and weekly newspapers, South African government reports, and reports by various anti-apartheid groups. When the office closed that afternoon, I left with my briefcase stuffed full of photocopies of materials directly pertinent to my investigation; for instance, an article from a South African newspaper praising LaRouche's economic theories; an article from another South African newspaper quoting a government commission as charging that the first newspaper was funded by BOSS (the South African secret police) and reports suggesting that BOSS propagandists had worked directly with the LaRouchians. I don't think I could have done too much better with help from the CIA—and this is only one of many such experiences I have had with the clippings files of nonprofit organizations.

# Appendix A ·

## Investigating Cults and Political Extremist Organizations

In recent years, two extremist traditions in America—that of the totalitarian and/or bigoted political sect, and that of the religious cult—have converged. Sects on both the right and the left have turned to cultist methods of recruitment and control (which fit perfectly with their totalitarian beliefs anyway). Religious cults, in turn, have gone political, attempting to link up with the mainstream right and bearing their own peculiar brand of totalitarianism.

## A.1  Where to Look First

Contact first the organizations that monitor extremism full-time. The Cult Awareness Network (CAN), based in Chicago, tracks religious and political cults of all stripes, as does (in the heart of cultland) the Commission on Cults and Missionaries of the Jewish Federation of Greater Los Angeles. The Anti-Defamation League of B'nai B'rith (ADL), with headquarters in New York, and the Simon Wiesenthal Center in Los Angeles monitor anti-Semitic groups of the right and left. The Atlanta-based Center for Democratic Renewal and the Montgomery, Alabama-based Klanwatch Project of the Southern Poverty Law Center track right-wing hate groups. Political Research Associates (PRA) in Cambridge, Massachusetts, reports on authoritarian tendencies in the mainstream right as well as the far right; it also has a unique expertise on political cults of both the right and left. For information on extremist groups from a conservative perspective, contact *Information Digest* in Baltimore.

Each of the above organizations issues newsletters and/or special reports and maintains extensive clippings files. CAN has files on over 1,300 cults and sects, and will mail you photocopies of selected file items. The Jewish Federation of Greater Los Angeles operates the Maynard Bernstein Resource Center on Cults, with files on several hundred groups, a library of videotapes (this is important since so much of the best reporting on cults is done by TV journalists), and a collection of legal briefs, depositions, etc., from cult-related lawsuits. The ADL's huge library of clippings covers the past fifty years of American extremism. ADL staffers will search these files for journalists and send them photocopies.

The monitoring organizations may have a file on the sect or cult you are researching but may not have any in-depth knowledge of it. Always ask them to refer you to someone who specializes in tracking the group in question.

## A.2 The Target Group's Own Propaganda

It is important to study back issues over a period of years of a group's newspapers and periodicals, as well as its pamphlets, internal strategy documents, etc. If not in the files of one of the monitoring organizations, these publications and documents can be found in various research libraries. The biggest collection of extremist newspapers and periodicals (about 7,500 separate titles) is at the State Historical Society of Wisconsin Library in Madison. Other notable collections on extremism are at New York University (the Tamiment Institute Library), Brown University (the Hall-Hoag Collection), University of Michigan at Ann Arbor (the Labadie Collection), and University of Kansas (Wilcox Collection). For details on these and many others collections on extremism, see the *Directory of Archives and Manuscript Repositories in the United States, Subject Collections,* and the *Directory of Special Libraries and Information Centers.*

If you want to see what is in the Wisconsin collection, consult *Periodicals and Newspapers Acquired by the State Historical Society of Wisconsin, July 1974 Through December 1985,* a 970-page annotated and cross-indexed list available in large research libraries. Publications from the Wisconsin collection that are available on microfilm can be obtained through interlibrary loan. If the particular publication you need is not in the Wisconsin collection, or if you want to find a backfile in a library in your own region, look in the *Union List of Serials* (serials before 1950) and *New Serial Titles* (after 1950). For other materials on a particular group—pamphlets, archival manuscripts, etc.—check the OCLC and RLIN databases (described in 11.2).

### Newspapers Or Newsletters
Although some extremist groups are supersecretive, most like to boast about their successes. Often their weekly or monthly newspaper or newsletter will present a detailed picture of their public activities. By checking out the newspaper's claims of success (e.g., asking a local board of election

to verify election returns), you can begin to get a picture of just how innocuous or dangerous a group really is. In addition, back issues can help you compile a roster of the group's members: those listed on the masthead as editorial or production staff members, those who sign articles (if they don't all use pen names), those reported as running for public office, those mentioned or quoted in articles about the group's demonstrations and conferences, those reported as being targets of police brutality or parental attempts at deprogramming, those named as the spouses of members who die (obituaries in these publications are always interesting), and those praised as Fundraiser of the Month.

## "Theoretical" Literature

Since the group's newspaper is aimed at the general public, the ideas therein are sometimes toned down or sanitized. Often a better picture of the group's thinking can be obtained from its theoretical journal—if such exists—or from books, pamphlets, articles, or audio or video tapes of speeches by its leader/guru. Such materials may be advertised for sale in the group's newspaper. Although most journalists are too lazy to search out and carefully read the turgid and often repulsive writings of sect or cult leaders, such documents are often the key to what the group is all about. To understand the Unification Church, you must read Rev. Moon's *Divine Principle*. To understand the Liberty Lobby, you must read Francis Parker Yockey's *Imperium*. To understand the Aryan Nations sect, you must read William Pierce's *The Turner Diaries*.

Many groups keep their core ideology secret from outsiders and only express it in unpublished internal documents. In some cases, fanatical books or pamphlets published years ago have been removed from public circulation to further a group's current deceptive tactics. Often the earlier documents can be obtained from former members or at a research library archival collection.

## Publication Indexes

*The Militant* and *Workers Vanguard* are among several leftist party newspapers that publish annual indexes. Many of the left-wing theoretical magazines are also indexed. At the opposite end of the spectrum is the annual index of *The Spotlight*. Note that both ultraright and ultraleft publications place a heavy emphasis on criticizing rival groups; you can therefore pick up much gossip by searching their indexes for articles attacking or exposing their principal rivals.

## Radio/TV Talk Show Appearances

Leaders of extremist groups pop up incessantly on radio talk shows. One of the main reasons they wage seemingly quixotic campaigns for public office is to wangle such appearances. Although no national index of talk-show guest appearances exists, you can find out when and where members of the group in question were on radio or TV by looking in back issues of its newspaper for announcements or articles regarding these forays. Often the talk show host (or his call-in questioners) will have baited or cajoled the extremist leader into making indiscreet remarks. The "air tapes" of local radio shows are kept for varying periods, depending on station policy (generally you should try to get the tape as quickly as possible). To obtain a cassette tape, call the station's program director. Local stations often provide these cassettes as a courtesy to journalists, while national network shows will sometimes send you a transcript.

If you are interested in an extremist leader's talk show appearances in a given city or region (but don't find any announcements in his or her newspaper), see the *Gale Directory of Publications and Broadcast Media* for a list of local stations. A few calls to station managers will elicit the names of local talk-show hosts who frequently invite extremists on the air. Some stations compile guest lists for their talk shows which are sent to local newspapers beforehand. If a station keeps a backfile of these guest lists you might search through them to find the date of Ned the Nazi's appearance.

### Extremist-Sponsored Radio and TV Programs

Some extremist groups have their own radio programs. Louis Farrakhan rose to prominence as a radio preacher in the Chicago area. The ultra-right Liberty Lobby produces a two-hour daily talk show, Radio Free America, which can be picked up by satellite dish owners. (The Lobby also runs the Sun Radio Network, which distributes Radio Free America and other questionable programs to 147 stations.)

Tom Metzger, head of the White Aryan Resistance (WAR) is among the extremists using public access cable TV to spread their ideas: Metzger's "Race and Reason" show appears on 31 public access channels across the country. According to a 1991 ADL report, cable TV hate shows are now being aired in 24 of the nation's 100 largest cable markets. The report lists 57 of these programs from Los Angeles to Boston.

### Video and Audio Cassette Propaganda

Video cassettes have become a popular method of propaganda among conspiracy-oriented rightist groups in the farm belt. Watching these videos may give you a better picture of a particular group's goals and methods than would its semi-literate newsletter.

You will find these video cassettes, as well as audio ones, advertised in ultra-right newsletters and mail-order catalogs. They may also be advertised on an extremist-sponsored radio or cable TV show.

Extremist videos often are taped at a rally addressed by the group's leader. By freezing the tape you can sometimes identify faces in the audience (including that of an investigative journalist who has infiltrated the meeting!). Note especially those people who share the podium with the leader.

### *"Hate Lines" and Other New Types Of Propaganda*

WAR and other extremist groups have hot-line (or "hate-line") telephone numbers in various cities. When you call, you hear a recorded message that is usually replaced once a week. The hate lines are often advertised in the newsletter of the sponsoring group. You can also get a list of the numbers from either the Center for Democratic Renewal or the ADL.

Another new form of propaganda is the computerized bulletin board. Several years ago, a hacker in Upper Manhattan obtained a password to a white supremacist bulletin board and downloaded the equivalent of hundreds of pages of computerized rantings. The results of this raid suggest that the computer keyboard has replaced the soapbox as the preferred means of neo-Nazi agitation.

## A.3 Defectors, Parents, and Other Special Sources

Extremist groups often have a high membership turnover. Some members quietly resign as a result of burn-out. Others are kidnapped by their parents and "deprogrammed." Still others form factions and/or concoct heretical position papers—they are then denounced by the group's leadership and expelled, if they don't walk out first.

If you can find these defectors, they may be useful in many ways. Besides giving you their first-person accounts, they may provide internal documents of the group and copies of their own position papers. They also may introduce you to fellow defectors and gather documents for you from defectors too paranoid for direct contact. They may even be in touch with someone still inside the organization who is on the verge of walking out and possibly can be persuaded to bring along a copy of the general ledger.

Defectors from religious and political cults may be contacted through CAN, which is linked to parent support groups and exit counselors around the country. If CAN or other monitoring organizations are unable to help you, you can search for defectors on your own: Go to back issues of the group's newspaper, and look for two things: first, accounts of factional strife that may include the names of dissidents; second, the sudden disappearance, after many years, of a featured byline or of a top editor's name from the masthead (this person may have been expelled for factionalism or may simply have dropped out). You might also check the annual 990 forms of the group's nonprofit entities for instances in which persons listed as officers in past years are no longer listed.

### *Splinter Groups and Longtime Rivals*

For obscure reasons, some extremist traditions are characterized by ferocious factionalism and frequent splits. Members will leave a particular group just to dig a new frog pond for themselves (i.e., the "new" Klan or the "new" Fourth International). Often the splinter group will start its own newsletter, which, for the first few issues at least, will be almost totally devoted to denouncing the parent group. You may first learn about such a publication when the newspaper of the parent organization issues a statement denouncing it. Often the splinterists will be eager to talk to a journalist from the outside world.

You also may find decades-old feuds between two fringe groups that regard each other as the ultimate in heresy or treachery. If the leader of Group A thinks he can use a journalist to make life miserable for the leader of Group B, he may talk quite frankly off the record. (In the 1930s and 40s, Stalinists in New York were always willing to feed journalists the latest "proof" that the Trotskyists were agents of Hitler.) Obviously you should triple-source anything these biased individuals tell you.

### Parents and Relatives Of Brainwashed Members

Some cult leaders encourage their followers to break off all ties with their parents. Other cult leaders encourage their disciples to borrow as much money as possible from the parents, and *then* break all ties. Even when the cult's attitude to parents is more relaxed, the parents go through hell when Johnny is only allowed to talk to them on the phone once a month. In an attempt to get a handle on the situation, they may hire a private investigator to compile a dossier on the cult leader. Or they may conduct their own investigation: visiting the cult's headquarters, collecting cult literature, and questioning their youngster closely during his or her infrequent visits home (these chats at the kitchen table can elicit a surprising amount of information). A Long Island mother in the mid-1980s let herself be recruited to a cult in an attempt to get her daughter out. Such parents can be of great help as you prepare your media expose, as can siblings and other relatives who may be every bit as upset as the parents are. (Note that some political and religious cults are now focusing on senior citizens—one thus gets a new scenario in which adult children are scheming to deprogram Mom.)

## A.4 Election Campaign Records

Although some extremist groups eschew electoral politics, the more sophisticated see it as a useful tool. Groups with purist leaders run candidates only under their own banner; others advocate a flexible "tripartite" strategy (run as a Democrat, a Republican, or openly as an ideological candidate depending on the circumstances). In 1986, an ultra-right party fielded over 500 candidates for public office. In 1988, a far-leftist party managed to get its presidential candidate on the ballot in all fifty states.

Because of state and federal election laws, such candidates cannot obtain ballot status without revealing considerable information about their group's finances, volunteer support networks, and subterranean political connections. By carefully studying the election filings of extremist party candidates (which are always public record information), you can get an excellent picture of who's who in the organization locally and nationally. The filings may also give you a window into the more violent and secretive groups that do not run candidates themselves but will sometimes support a more sedate ultra-rightist at the polls. For instance, many Posse Comitatus members gave money to LaRouchian and Populist Party candidates in the mid-1980s, and many Klansmen and Christian Patriots backed David Duke's Louisiana campaign for the U.S. Senate in 1990.

### Nominating Petitions
Candidates for local, state, and federal office must file nominating petitions in order to enter a major party primary or get on the ballot as an independent or minor-party candidate in the general election. Depending on which office the candidate is running for, thousands of signatures (with addresses) may be required. Party or cult leaders, themselves aloof from the work of petitioning, may sign to help fill up a sheet, thus revealing the address of their secret bunker. You might browse through the nominating petitions at the Board of Elections to see how many addresses of known cult members you can find—this could give you a unique picture of their communal living arrangements.

Most of the other signatures will be useless to an investigator, since the signers will have no real connection to the party or candidate—they may have thought they were signing for George Bush. The chief importance of the petitions is that each sheet must be signed at the bottom by the person who gathered the signatures—the petition "witness." Examining the witness signatures enables you to compile a list of those party supporters who are sufficiently committed that they will scurry around the streets or ring doorbells to collect signatures. If the local Populist Party turned in 200 petition sheets signed by a total of fourteen petition gatherers (say, for the Populist candidate for Congress), you may have gained thereby a list of virtually the entire active membership in your area.

### Committees On Vacancies
In New York, each candidate in a primary election must appoint a committee on vacancies—a committee of supporters who would appoint a person to take the candidate's place if the latter should die between the time of filing and the primary. Likewise, each independent or minor-party candidate must appoint a committee to receive notices. Similar requirements exist in other states. The names and addresses of these committee members will either be on the nominating petition itself or be available from the board of elections.

### Presidential Campaigns
If the leader of an extremist group decides to run for president in a Democratic or Republican primary, he or she must (depending on the

laws of the particular state) file delegate slates for the county caucuses, the state party convention, or the national party convention. And if the candidate chooses to run in November on a minor-party ticket, he or she must file a list of presidential electors for each state in which he or she qualifies for the ballot. These must be registered voters of the state, and their number must correspond with the number of presidential electors assigned to the state under the Electoral College System. (Thus, in 1984, the candidate of the Populist Party had to file the names and addresses of eleven electors in Tennessee.) Extremist presidential candidates rarely get on the ballot in a majority of states, but even an effort in a few states may bring out of the woodwork a hundred or more supporters whose names you might not have learned otherwise.

## Campaign Financial Reports

The campaign committees of all candidates for federal elected office must file detailed reports with the Federal Election Commission (FEC) if they raise or spend over $5,000. These reports provide an accounting of receipts, disbursements, and campaign debts. Presidential candidates applying for matching funds must supply additional information about contributions, sometimes totaling thousands of pages. And PACs must file reports like those of the campaign committees. Included in these filings will be complete lists of contributors and amounts contributed (for any amount over $200), lenders and amounts lent, and also information about vendors who extend credit to the campaign in a manner that suggests they are really closet supporters (like the local barbecue restaurant that extended credit to Ned the Nazi for a campaign dinner on its premises even though Ned still owed them for similar services during a campaign two years earlier).

Less-detailed statements may be required on the state and local level. Generally, state candidates will file with the state elections board and city candidates with the city elections board. If one takes all the filings in recent years by ultraright candidates for local office as well as those for the U.S. Senate, Congress, and the Presidency, one is looking at lists of well over 100,000 contributors. All in all, these campaign finance reports provide a "who's who" of local and national extremist support networks in America.

The easiest way to access this who's who is via the FEC's online Contributor Search System. Do you want to know if John Doe of Midtown, Iowa, has contributed to any white supremacist candidates for federal office? The Contributor Search System will tell you all reported contributions of more than $200 by Doe to any federal candidate, extremist or not. Do you want an idea of who might be part of the white supremacist network in Louisiana? The FEC's Index G will tell you the name, city, and occupation of all reported Louisiana contributors of $500 or more to David Duke's 1988 presidential campaign.

Election campaign filings can also help you estimate the types of people an extremist candidate is reaching. For instance, you may want to know if a certain race-baiting presidential candidate in a three-piece suit is getting support from people of potential influence in the community—or if he is merely picking up the eccentrics. In part you can judge this by searching the FEC databases, which give the occupation and place of employment of contributors.

## A.5 Newspaper Circulation Figures

The circulation figures of an extremist organization's newspaper (and especially the number of subscribers) may be useful in measuring the organization's growth or decline. You can find this information on the second-class mailing permit statement of ownership that the publisher must file with the U.S. Post Office each year. The statement must include the publisher's name and address, the names of corporate officers, and detailed circulation information (number of copies printed in the latest issue; average number printed in the past year; and a breakdown of latest and average figures for subscription sales, other types of sales, and giveaways). This statement must be published annually in the publication. If you can't find the issue in which it appeared, you can obtain a copy through an FOIA request to the central post office of the city in which the publication is located.

These are usually unaudited figures, and many extremist groups doubtless take the opportunity to exaggerate their support. One way to get around this is to call the printer and, via an appropriate pretext, induce him to tell you what the press run really is. But even the press run may be misleading; the extremist group in question may be printing tens of thousands of copies to pass out for free at demonstrations or simply to leave in stacks at supermarkets. A more meaningful measure of public support is the number of subscribers. If the subscriptions are being handled by a commercial jobber you might find out the number via a pretext phone call, but this figure should also not be taken at face value: The LaRouche organization, for instance, sells subscriptions at airports via high-pressure tactics and also induces wealthy supporters to each buy hundreds of annual "gift" subscriptions to send unsolicited to libraries and public officials. A very large percentage of the airport subscriptions (sold to people who had no idea what they were buying) are not renewed; and most of the gift subscriptions go into the waste basket. A true picture of any extremist publication's support would include the total number of non-gift subscriptions, and also the subscription renewal rate.

## A.6 Piercing the Corporate Veil

The larger cults and extremist groups often operate a tightly woven network of corporations, unincorporated entities, and individual d/b/a's, with the income cavalierly treated according to the Three Musketeers principle (all for one, and one for all). The network may include businesses that function in the normal work-a-day world (e.g., Rev. Moon's fishing fleets), telemarketing and other fundraising entities, a mail-order book service and/or a chain of bookstores, an in-house typesetting or printing business, assorted publications, one or more tax-exempt foundations, a summer retreat for cult leaders, a video production facility, and cash businesses such as drug smuggling. Often the network will be used to move around money as part of check kiting, tax evasion, and money-laundering schemes. (You move the money from account to account, and then it disappears either into the leader's pocket or into a slush fund for the stockpiling of automatic weapons.)

The various components of an extremist network leave a paper trail, including incorporation papers, annual corporate and not-for-profit filings, tax liens, UCC-11 statements, etc. Some of the entities will be listed in court indexes as plaintiffs or defendants in lawsuits. Some will also show up in FEC and state and city election campaign filings as vendors of services to the group's candidates.

The trick is to show how these entities fit together and how they are centrally controlled by the cult or party leadership. If you succeed, you may open the path to a RICO suit, IRS tax fraud case, or successful collection efforts by scam victims. Of course, if you are a citizen investigator you will not have access to the confidential bank records that the IRS can get. But you can accumulate enough evidence to start the ball rolling. Look first for an overlapping of officers and staff (i.e., the same people heading up or working for different entities at the same time), a sharing of facilities (same office or phone), and a sharing of vendors, attorneys, etc. Here is a check list I once used in examining Lyndon LaRouche's corporate veil:

- overlap of incorporators of business and nonprofit entities

- overlap of officers and/or principals of various entities (as reflected in annual state corporate filings, federal 990 forms, and second-class mailing permit statements)

- use of the same attorney by different entities during the same period— and the simultaneous shift by all entities to another attorney

- use of payroll checks drawn on the same payroll account and/or processed by the same payroll company

- use of the same petty cash fund and vouchers

- use of the same in-house bookkeeping facilities and the same accountant

- use of the same offices, phone number, fax number, P.O. box, mail drop, or answering service

- use of the same postal meter machine, imprint permits, etc.

- use of a single internal phone directory for the employees of all entities (defectors can help you with this)

- use of the same in-house notary

- use of the same process server (often a party or cult member) by different entities in different lawsuits

- co-listings in a building lobby directory and floor directory, and on the door of the office suite

- listings as co-plaintiffs or co-defendants in lawsuits

- listings as co-debtors in judgment books or UCC filings

- co-occupancy of various properties owned or rented by a single entity (if this involves subleasing, it can often be confirmed through the building's landlord).

It is important to establish each of the overlaps not just at a single moment in time but over an extended period. This may mean drawing diagrams and charts and/or doing some cross-filing on your computer. Let's say that Entity A and Entity B of the Aryan Peoples Party (APP) have shared a phone and office for the last four years on Elm Street and for the previous two years on Bates Street. Entity C has only shared the phone and office for the past three years on Elm Street and never on Bates Street. However, Entity C and Entity A have had the same treasurer for the past four years, while Entity B had the same president (but never the same secretary) as Entity A until three years ago. But Entity B's new president is, in turn, listed as the incorporater, six years ago, of Entity C. In a tightly organized extremist group, where the leader only trusts a few close lieutenants, you may find precisely this kind of linkage.

One must also show that underneath the permutations, the common thread is the APP and that the APP in fact controls the various entities even though in a legal sense neither the APP nor its chairman has any ownership interest in them. Here are just a few of the things you can check:

- Have any of the entities ever operated out of the APP's national head-quarters (or vice-versa)?

- Have any of the entities shared office space with an APP branch office?

- Do officers of the entities write for or serve on the editorial board of the APP publications?

- Are officers of the entities identified as APP members or APP executive committee members in APP newspaper and periodical articles?

- Do internal APP documents or minutes discuss the affairs of the entities in a way that suggests the APP actually controls them?

- Have any entity officers ever identified themselves as APP members in an affidavit, a deposition, or trial testimony?

- Have any entity officers run for public office as APP candidates?

- Have any been listed in FEC, state, or local campaign filings as having donated money to APP campaigns (and have any other entity employees made such donations)?

- Have any entity officers witnessed petitions for an APP campaign or been designated as members of a committee to receive notices for an APP campaign?

- Have any APP campaign committees in past years shared office space with any of the entities?

- Have any of the entities served as vendors (e.g., printers or typographers) for APP campaigns? Did they extend credit to any past APP campaign committee to an extent that suggests the work performed was really an unreported contribution-in-kind? Do the latest FEC filings of such campaign committees show that the debt has still not been paid?

## A.7 Freedom of Information Laws

In past decades, the FBI obsessively monitored just about every extremist group in America. To obtain the FBI's file on one of these groups under the Freedom of Information Act usually takes a year or more because of the backlog of requests. However, a vast amount of material has already been released under the FOIA, and this may include your subject organization (OR the parent organization from which it split). The Reading Room at FBI headquarters has a vast collection of preprocessed FOIA documents. The list of topics, including organizations and deceased individuals, and the total number of pages available on each topic, is available from the FBI's FOIA unit. In addition, you can consult the Declassified Documents Reference System, a subscription service for libraries that includes an index, abstracts, and microfiche copies of released documents, among them many FBI documents. The National Security Archive in Washington, D.C., includes in its collection of mostly overseas-related documents about 1,000 FBI files, mostly regarding surveillance of anti-Vietnam war activists. Released FBI files of special historical value are published on microfilm by Scholarly Resources, Inc. in Wilmington, Delaware. Political Research Associates (see above) offers photocopies of COINTELPRO files on the New Left and black nationalist groups sorted by city (over 40,000 pages in all). In addition, documents released only to the requesting individuals (mostly their own FBI files) are often entered as exhibits in civil liberties lawsuits against the government, and the plaintiffs frequently obtain still more documents via pretrial discovery.

If you are on a short-range project, I would advise making an FOIA request anyway. Your project may turn out to last longer than you think, and the extremist group in question is probably not going to disappear during the period you wait for the documents. I would also advise making requests from agencies other than the FBI, including the Bureau of Alcohol, Tobacco, and Firearms, which tracks stolen explosives and illegal stockpiles of automatic weapons.

In any request for FBI files, specify that you want field office as well as headquarters files on your target organization. Make separate requests for the files on each of the organization's front groups and spinoffs. Be sure to request files for defunct front groups—and also for any defunct parent organization—in case you have trouble getting the files on the currently existing organizations. In addition, you should request the personal files of every member of the target organization who has died, and get as many defectors as possible to request their own files. These requests for personal files may produce material that is less heavily redacted than the organizational files; also, the Freedom of Information official working on one request may release a document that another official would not let pass (and vice-versa). If you take the time to collect and collate all this material, you will have an astonishingly detailed record of your target group's evolution.

An excellent handbook, *Are You Now or Have You Ever Been in the FBI Files: How to Secure and Interpret Your FBI Files,* will tell you everything else you need to know on FOIA requests; see bibliography.

Much of the FBI's historical information about extremist groups and their leaders up through the early 1970s, is duplicated in the Church League of America's files. This now defunct conservative organization donated its files on leftists and on far-right hate groups (7 million index cards and 200 file cabinets) to Jerry Falwell's Liberty University in Lynchburg, Virginia, where they are currently unavailable to the public. Even if they should become accessible, much of the activity they document would not be regarded as subversive by today's standards (the same of course is true of many of the FBI's files).

## A.8 Monitoring and Surveillance

In monitoring far-right groups on an ongoing basis, there are a variety of tactics from which to choose. These vary in the degree of risk involved, and the riskier ones should not be undertaken lightly.

First, you can simply take out a subscription to the Aryan Peoples Party's newspaper and also get on the mailing list to receive its fundraising letters and mail-order book, pamphlet, and cassette sales literature. In doing this, you might be tempted to rent a post-office box and receive your subscription under the name "Ragnar Northman" or "Horst Wessel." Be aware, however, that the extremist group can easily find out who you are by calling up the post office and claiming that the boxholder of Box 662 sold them something through the mail and it never arrived—often the postal clerk will give them the box holder's name and address on the spot.

Second, you can open an account with the FEC by paying a deposit, and have the FEC mail you all reports of APP-linked PACs and candidate committees as soon as filed (including reports from no-longer-active committees from past elections that still must file regarding their outstanding debts and obligations).

Third, you can tape regularly the APP leader's farm belt radio show, his public-access cable TV show, and his telephone hate tapes in various cities.

Fourth, you can join the APP's computer bulletin board network and download their demented discussions (but beware of computer viruses).

Fifth, you can get on the APP's special mailing list for upcoming rallies and Oktoberfests, and then show up to count heads, gather auto license plate numbers (see 11.5), and tape the speeches.

Sixth, you can take the ultimate step and infiltrate the ranks of the racial comrades, but that is a tactic outside the scope of this book.

# APPENDIX B

## Journals
*The IRE Journal* (bimonthly), 100 Neff Hall, University of Missouri, Columbia, MO 65211. Contains state-of-the-art trade tips unavailable anywhere else. Complete backfiles are available at a reasonable price. Be sure you get a copy of IRE's 1989 Membership Directory, which includes a cumulative index to the IRE Journal backfiles.

*The Legal Investigator* (quarterly), 3304 Crescent Drive, Des Moines, IA 50312.

*Special Libraries* (quarterly), 1700 18th Street N.W., Washington, DC 20009.

## Newsletters
*CD-ROM Databases,* WV Publishing Company, Box 138, Babson Park, Boston, MA 02157.

*Lesko's Info-Power Newsletter* (formerly *Data Informer*), Information USA, P.O. Box E, Kensington, MD 20895.

*Privacy Journal,* P.O. Box 28577, Providence, RI 02908. Also sells a back-file (1974-1989) index.

*Private Investigators' Connection,* Thomas Publications, P.O. Box 33244, Austin, TX 78764.

*Uplink,* Missouri Institute for Computer-Assisted Reporting, 120 Neff Hall, University of Missouri, Columbia, MO 65211.

## Reference Book, Microform, and CD-ROM Publishers
Commerce Clearing House, Inc., 2700 Lake Cook Road, Riverwoods, IL 60015.

Dow Jones-Irwin, 1818 Ridge Road, Homewood, IL 60430.

Dun & Bradstreet, 1 Penn Plaza, New York, NY 10119.

Gale Research Inc., 835 Penobscot Building, Detroit, MI 48226.

H.W. Wilson Company, 950 University Avenue, Bronx, NY 10452.

Marquis Who's Who, 3002 Glenview Road, Wilmette, IL 60091.

Meckler Corporation, 11 Ferry Lane West, Westport, CT 06880.

Moody's Investors Services, 99 Church Street, New York, NY 10007.

NewsBank, Inc., 58 Pine Street, New Canaan, CT 06840.

TRW REDI Property Data, 3610 Central Avenue, Suite 500, Riverside, CA 92506.

R.R. Bowker, 245 West 17 Street, New York, NY 10011.

Standard & Poor's Corporation, 25 Broadway, New York, NY 10004.

University Microfilms International, 300 North Zeeb Road, Ann Arbor, MI 48106.

West Publishing, 50 West Kellogg Boulevard, Saint Paul, MN 55102.

NOTE: Many of the above also produce online databases.

### Database Vendors

DataTimes, 14000 Quail Springs Parkway, Oklahoma City, OK 73134.

Dialog Information Services, Inc., 3460 Hillview Avenue, Palo Alto, CA 94304.

Information Access Company, 362 Lakeside Drive, Foster City, CA 94404.

Maxwell Online, Inc. (ORBIT Search Service), 8000 Westpark Drive, McLean, VA 22102.

Mead Data Central (LEXIS and NEXIS), 9443 Springboro Pike, P.O. Box 933, Dayton, OH 45401.

NewsNet Inc., 945 Haverford Road, Bryn Mawr, PA 19010.

UMI/Data Courier, 620 South Third Street, Louisville, KY 40202.

VU/TEXT Information Services, Inc., 325 Chestnut Street, Philadelphia, PA 19106.

### Gateway Services

CompuServe, 5000 Arlington Centre Boulevard, P.O. Box 20212, Columbus, OH 43220.

Sprint Gateways, P.O. Box 7910, Shawnee Mission, KS 66207.

Telebase Systems/EasyNet, 435 Devon Park Drive, Wayne, PA 19087.

The Source, 1616 Anderson Road, McLean, VA 22102.

### Specialized Investigative and Public-Records Gateways and Information Brokers

Datafax Information Services, P.O. Box 33244, Austin, TX 78764 (this firm markets the National Credit Information Network as well as other services—a kind of "broker of brokers").

Information America, One Georgia Center, 600 West Peachtree Street N.W., Atlanta, GA 30308.

Metronet, 360 East 22 Street, Lombard IL 60148.

National Credit Information Network, W.D.I.A. Corporation, 7721 Hamilton Avenue, Cincinnati, OH 45231 (this firm probably offers the broadest access to investigative databases—including all major credit networks).

Prentice Hall Online, 1900 East 4th Street, Santa Ana, CA 92705.
Rapid Information Services, 1710 West Roosevelt, Phoenix, AZ 85007.
Super Bureau Inc., P.O Box 368, Campbell, CA 95008.
Worldwide Tracer's Service, P.O. Box 48, Elmhurst, IL 60126.

### *How-To Publishers' Catalogs*
Eden Press, P.O. Box 8410, Fountain Valley, CA 92728.
Loompanics Unlimited, P.O. Box 1197, Port Townsend, WA 98368.
Thomas Publications, P.O. Box 33244, Austin, TX 78764 (ask for "The
    P.I. Catalog").

# BIBLIOGRAPHY

### Investigative Techniques—General

Anderson, David and Peter Benjaminson. *Investigative Reporting*, 2d ed., Ames: Iowa State University Press, 1990.

Blye, Irwin and Ardy Friedberg. *Secrets of a Private Eye*, New York: Henry Holt, 1987.

Downie, Leonard, Jr. *The New Muckrakers*, New York: New American Library, 1978.

Golec, Anthony M. *Techniques of Legal Investigation*, 2d ed., Springfield, Ill.: Charles C. Thomas, 1985.

Harry, M. *The Muckraker's Manual: Handbook for Investigative Reporters*, Port Townsend, Wash.: Loompanics Unlimited, 1984.

Mollenhoff, Clark R. *Investigative Reporting: From Courthouse to White House*, New York: Macmillan, 1981.

Pawlick, Thomas. *Investigative Reporting: A Casebook*, New York: Richards Rosen Press, 1982.

Pileggi, Nicholas. *Blye, Private Eye*, Chicago: Playboy Press, 1976.

Rose, Louis J. *How to Investigate Your Friends and Enemies*, St. Louis: Albion Press, 1981. Especially valuable are Chapter IV ("Investigating Real Estate") and Chapter V ("Finding the Hidden Owners").

Thompson, Josiah. *Gumshoe*, New York: Fawcett Books, 1989.

Ullmann, John and Jan Colbert, eds. *The Reporter's Handbook: An Investigator's Guide to Documents and Techniques*, 2d ed., New York: St. Martin's Press, 1991. This is an indispensable work for any investigator or journalist.

Weberman, A.J. *My Life in Garbology*, New York: Stonehill Publishing Company, 1980.

Weinberg, Steve. *Trade Secrets of Washington Journalists*, Washington, D.C.: Acropolis Press, 1981.

Williams, Paul N. *Investigative Reporting and Editing*, Englewood Cliffs, N.J.: Prentice-Hall, 1978.

## Examples Of Investigative Journalism/Research

Appalachian Land Ownership Task Force. *Who Owns Appalachia?*, Lexington: University Press of Kentucky, 1983.

Bamford, James. *The Puzzle Palace,* New York: Penguin Books, 1983.

Bellant, Russ. *Old Nazis, the New Right, and the Republican Party,* Boston: South End Press, 1991.

Bernstein, Carl and Bob Woodward. *All the President's Men,* New York: Warner Books, 1976.

Colbert, Jan and Steve Weinberg, eds. *The Investigative Journalist's Morgue,* Columbia, Mo.: Investigative Reporters and Editors, 1990. An index to thousands of stories and series from IRE's files.

Goodwin, Jacob. *Brotherhood of Arms: General Dynamics and the Business of Defending America,* New York: Times Books, 1985.

Goulden, Joseph C. *The Benchwarmers: The Private World of the Powerful Federal Judges,* New York: Weybright and Talley, 1974.

Hunt, Linda. *Secret Agenda: The United States Government, Nazi Scientists, and Project Paperclip, 1944-1990,* New York: St. Martin's Press, 1991.

King, Dennis. *Lyndon LaRouche and the New American Fascism,* New York: Doubleday, 1989.

Lucian of Samosata. "Alexander or The Bogus Oracle," in Lucian, *Satirical Sketches,* trans. by Paul Turner, Baltimore: Penguin Books, 1961. This sketch, written in the second century A.D., is perhaps the world's first example of investigative reporting.

Mitford, Jessica. *Poison Penmanship: The Gentle Art of Muckraking,* New York: Noonday Press/Farrar, Straus and Giroux, 1988. Includes her celebrated expose of the Famous Writers School, "Let Us Now Appraise Famous Writers."

Moldea, Dan. *The Hoffa Wars,* New York: Paddington Press, 1978.

Naylor, R.T. *Hot Money and the Politics of Debt,* New York: Simon and Schuster, 1987.

Shilts, Randy. *And the Band Played On,* New York: St. Martin's Press, 1987.

Weinberg, Steve. *Armand Hammer: The Untold Story,* Boston: Little, Brown & Co., 1989.

Weir, David and Dan Noyes, eds. *Raising Hell: How the Center for Investigative Reporting Gets the Story,* Menlo Park, Calif.: Addison-Wesley, 1983. Case studies in investigative journalism.

Winks, Robin W., ed. *The Historian as Detective: Essays on Evidence,* New York: Harper & Row, 1969.

## Finding People

Askin, Jayne with Bob Oskam. *Search: A Handbook for Adoptees and Birthparents,* New York: Harper & Row, 1982.

Gunderson, Ted L. with Roger McGovern. *How to Locate Anyone Anywhere Without Leaving Home,* New York: E.P. Dutton, 1989.

Johnson, Richard S. *How to Locate Anyone Who Is or Has Been in the Military: Armed Forces Locator Directory,* Ft. Sam Houston, Tex.: Military Information Enterprises, 1990.

Thomas, Ralph D. *Advanced Skip Tracing Techniques,* Austin, Tex.: Thomas Publications, 1988.

Thomas, Ralph D. *Electronic Skip Trace Techniques,* Austin, Tex.: Thomas Publications, 1989.

Thomas, Ralph D. *How to Find Anyone Anywhere,* Austin, Tex.: Thomas Publications, 1983.

### Database Searching

Glossbrenner, Alfred. *How to Look It Up Online,* New York: St. Martin's Press, 1987.

Thomas, Ralph D. et al. *How to Investigate by Computer: 1990,* Austin, Tex.: Thomas Publications, 1989.

### Interviewing

Brady, John. *The Craft of Interviewing,* New York: Vintage Books, 1977.

Metzler, Ken. *Creative Interviewing,* Englewood Cliffs, N.J.: Prentice-Hall, 1977.

### Library Research and Government Resources

Berkman, Robert I. *Find It Fast: How to Uncover Expert Information on Any Subject,* New York: Harper & Row, 1990.

Lesko, Matthew. *Lesko's Info-Power,* Kensington, Md.: Information USA, 1990. (This is an updated and reorganized version of the same author's *Information U.S.A.,* New York: Penguin Books, 1986.)

Morehead, Joe. *Introduction to United States Public Documents,* 3d ed., Littleton, Colo.: Libraries Unlimited, 1983.

Todd, Alden. *Finding Facts Fast,* Berkeley, Calif.: Ten Speed Press, 1979.

### Courts and The Law

Denniston, Lyle W. *The Reporter and the Law: Techniques of Covering the Courts,* New York: Hastings House, 1980.

Wren, Christopher G. and Jill Robinson Wren. *The Legal Research Manual: A Game Plan for Legal Research and Analysis,* 2d ed., Madison, Wis.: Adams & Ambrose Publishing, 1989.

### Public and Private Records

Carroll, John M. *Confidential Information Sources: Public & Private,* Stoneham, Mass.: Butterworth Publications, 1975.

Murray, Thomson C. *License Plate Code Book,* Mill Neck, N.Y.: Interstate Directory Publishing Co., 1985.

Murphy, Harry J. *Where's What: Sources of Information for Federal Investigators,* New York: Warner Books, 1976. Originally a CIA manual, this 452-page book deals with restricted as well as open files.

National Employment Screening Services. *The Guide to Background Investigations,* 4th ed., Tulsa, Okla.: Source Publications, 1990.

National Employment Screening Services. "Social Security Numbers Guide," pamphlet, Tulsa, Okla.: Source Publications, 1991.

Smith, Robert Ellis. *Compilation of State and Federal Privacy Laws,* 8th ed., Providence, R.I.: Privacy Journal, 1990.

Thomas, Ralph D. *Investigator's 1990 State Records Access Directory,* Austin, Tex.: Thomas Publications, 1989.

U.S. Dept. of Health and Human Services. "Where To Write for Vital Records: Births, Deaths, Marriages, and Divorces," pamphlet, Washington, D.C.: U.S. Government Printing Office, 1990.

## Local Investigative Guidebooks

Davis, Aurora E., ed. *Access to Public Information: A Resource Guide to Government in Columbia and Boone County, Missouri,* rev. ed., Columbia, Mo.: Freedom of Information Center, 1990.

Dolan, John P., Jr. and Lisa Lacher. *Guide to Public Records of Iowa Counties,* Des Moines: Iowa Title Company, 1987.

Jeffres, Leo W. et al. *Cleveland State University Journalists' Handbook Series.* Vol. I: *News Strategies.* Volume II: *Public Records Guide* (includes Cuyahoga County, City of Cleveland, and regional, state, and other offices). Cleveland: Communication Research Center of Cleveland State University, 1988.

Kronman, Barbara. *Guide to New York City Public Records,* 4th ed., New York: Public Interest Clearinghouse, 1991.

Trzyna, Thaddeus C. *California Handbook: A Comprehensive Guide to Sources of Information and Action,* 6th ed., Sacramento: California Institute of Public Affairs, 1990.

## Freedom of Information Act

Buitrago, Ann Mari and Leon Andrew Immerman. *Are You Now Or Have You Ever Been in the FBI Files: How to Secure and Interpret Your FBI Files,* New York: Grove Press, 1981. Explains the terminology, abbreviations, acronyms, etc. used in FBI documents, plus the organizational structure and office procedures of the FBI in past years.

Daugherty, Rebecca, ed. "How to Use the Federal FOIA Act," 6th ed., pamphlet, Washington, D.C.: Reporters Committee for Freedom of the Press, 1987.

House Committee on Government Operations. "A Citizen's Guide on Using the Freedom of Information Act and the Privacy Act of 1974 to Request Government Records," pamphlet, Washington, D.C.: Government Printing Office, 1991.

Marwick, Christine M. *Your Right to Government Information,* 2d ed., Carbondale: Southern Illinois University Press, 1986.

## Business Research

Community Press Features. *Open the Books: How to Research a Corporation,* Cambridge, Mass.: Urban Planning Aid, Inc., 1974. Valuable tips on researching local businesses and estimating the profits of small, privately held corporations.

Daniells, Lorna. *Business Information Sources,* rev. ed., Berkeley: University of California Press, 1985. A thorough guide, highly recommended.

Fuld, Leonard. *Competitor Intelligence,* New York: John Wiley, 1985.

Pitt, Harvey L. and Herbert M. Wachtell, eds. *Preserving Corporate Confidentiality in Legal Proceedings,* New York: Harcourt Brace Jovanovich, 1980.

Tracy, John A. *How to Read a Financial Report: Wringing Cash Flow and Other Vital Signs Out of the Numbers,* 3d ed., New York: John Wiley, 1989.

## White-Collar Crime

Clinard, Marshall B., ed. *Illegal Corporate Behavior,* Washington, D.C.: Government Printing Office, 1979.

Edelhertz, Herbert. *The Investigation of White-Collar Crime,* Washington, D.C.: Government Printing Office, 1977. See especially Appendix D, "The Seventh Basic Investigative Technique: Analyzing Financial Transactions in the Investigation of Organized Crime and White Collar Crime Targets."

Gregg, John. "How to Launder Money," pamphlet, Port Townsend, Wash.: Loompanics Unlimited, 1982.

Hoy, Michael. *Directory of U.S. Mail Drops,* Port Townsend, Wash.: Loompanics Unlimited, 1987.

*The Paper Trip I* and *The Paper Trip II,* rev. eds., Fountain Valley, Calif.: Eden Press, 1971.